# Biomechanics—Structures and Systems

# The Practical Approach Series

SERIES EDITORS

**D. RICKWOOD**
*Department of Biology, University of Essex*
*Wivenhoe Park, Colchester, Essex CO4 3SQ, UK*

**B. D. HAMES**
*Department of Biochemistry and Molecular Biology,*
*University of Leeds, Leeds LS2 9JT, UK*

Affinity Chromatography

Anaerobic Microbiology

Animal Cell Culture (2nd edition)

Animal Virus Pathogenesis

Antibodies I and II

Biochemical Toxicology

Biological Membranes

Biomechanics—Materials

Biomechanics—Structures
and Systems

Biosensors

Carbohydrate Analysis

Cell Growth and Division

Cellular Calcium

Cellular Neurobiology

Centrifugation (2nd edition)

Clinical Immunology

Computers in Microbiology

Crystallization of Nucleic Acids
and Proteins

Cytokines

The Cytoskeleton

Diagnostic Molecular Pathology
I and II

Directed Mutagenesis

DNA Cloning I, II, and III

Drosophila

Electron Microscopy in Biology

Electron Microscopy in
Molecular Biology

Enzyme Assays

Essential Molecular Biology I
and II

Fermentation

Flow Cytometry

Gel Electrophoresis of Nucleic
Acids (2nd edition)

Gel Electrophoresis of Proteins
(2nd edition)

Genome Analysis

HPLC of Macromolecules

HPLC of Small Molecules

Human Cytogenetics I and II
(2nd edition)

Human Genetic Diseases

Immobilised Cells and
Enzymes

Iodinated Density Gradient
Media

Light Microscopy in Biology

# Biomechanics— Structures and Systems

## A Practical Approach

Edited by
A. A. BIEWENER

*Department of Organismal Biology
and Anatomy,
The University of Chicago,
Chicago, USA*

IRL PRESS
—at—
OXFORD UNIVERSITY PRESS
Oxford New York Tokyo

*This book has been printed digitally and produced in a standard specification in order to ensure its continuing availability*

# OXFORD
UNIVERSITY PRESS

Great Clarendon Street, Oxford OX2 6DP

Oxford University Press is a department of the University of Oxford.
It furthers the University's objective of excellence in research, scholarship,
and education by publishing worldwide in

Oxford New York

Auckland Bangkok Buenos Aires Cape Town Chennai
Dar es Salaam Delhi Hong Kong Istanbul Karachi Kolkata
Kuala Lumpur Madrid Melbourne Mexico City Mumbai Nairobi
São Paulo Shanghai Taipei Tokyo Toronto

Oxford is a registered trade mark of Oxford University Press
in the UK and in certain other countries

Published in the United States
by Oxford University Press Inc., New York

ISBN 0-19-963267-7

Printed in Great Britain by

Antony Rowe Ltd., Eastbourne

# Preface

This book presents techniques and methods used to study a range of topics concerning the mechanical design and function of organisms at structural levels of their organization. In a companion volume (*Biomechanics—materials: a practical approach*, ed. J. F. Vincent, Oxford University Press), methods for investigating the mechanical properties of biomaterials are presented. Given the hierarchical and functionally integrative nature of organisms, studies of the mechanics of their structures must necessarily build on studies of the material properties of their tissues in order to understand how structural design yields higher level mechanical function. Consequently, some overlap between the two books is unavoidable. Questions of mechanical design and function, in turn, ultimately relate to organismal performance and the interaction of organisms with their physical environment. Much of this book focuses on musculoskeletal biomechanics, in which the design of muscular and skeletal elements is evaluated in the context of physical activities, such as running, flying, or feeding. Similarly, the mechanical design of circulatory systems (Chapter 10) concerns transport functions inside the organism related to carrying out these activities.

Although the main emphasis of the book is at an 'organismal' structural level, it should be noted that recent attempts to explore the biomechanics of cells and tissues make these promising areas of future research as well, particularly with regard to our understanding of mechanisms of signal transduction and the adaptive responses of biological tissues to mechanical stimuli. The book also maintains a strong comparative thread in most of the topics that are covered. It is hoped that for readers unfamiliar with the 'comparative method', the importance of this approach will emerge, not only for appreciating the intrinsic value of the organisms themselves, and their own phylogenetic history, but also in terms of its ability to discern basic patterns and principles of biological structure and mechanical function that have general application to other fields.

While the intent of the book is to present 'practical' approaches for studying the structural biomechanics of organisms, each chapter also includes conceptual and theoretical discussions as a background to the techniques and methods covered. In fact, in two of the chapters (Chapter 5, on the aerodynamics of flight; and Chapter 11, on the hydrodynamics of animal movement), concept and theory take precedence over technique. By doing so, our hope is to provide not only a useful 'guide' but also to convey some of the fascination, excitement, and importance of the topics that we study. The reader should recognize that most chapters serve only as a brief introduction to the techniques and methods presented. In many cases, additional reading is recommended; and in all cases, 'hands-on' experience is essential for achieving the expertise needed to use and understand these approaches.

As always, the need to balance breadth against the details of presentation within a finite space means that some useful techniques and interesting organisms have been left out. This is a book about living animals, and hence reflects my own research bias. Notably, the biomechanics of plants and fossil organisms are absent. The experimental approaches presented here, however, are not limited by the bounds of the topics that we have space for, or the fields that these topics represent. Hopefully, the biomedical engineer will gain insight into the utility of comparative biology, while the comparative biologist will appreciate the application of engineering mechanics to solve problems in physical biology. In the end, the book's success will depend not only on its usefulness as a guide, but its ability to foster the reader's interest in exploring new areas of biomechanics research at a structural level.

*Chicago*                                                                                          ANDREW A. BIEWENER
September 1991

# Contents

## 3. Force platform and kinematic analysis 45
*Andrew A. Biewener and Robert J. Full*

## 4. Mechanical work in terrestrial locomotion 75
*R. Blickhan and Robert J. Full*

# 5. Aerodynamics of flight 97

*Robert Dudley*

# 6. In vivo measurement of bone strain and tendon force 123

*Andrew A. Biewener*

# 7. Finite element analysis in biomechanics   149

*Gary S. Beaupré and Dennis R. Carter*

# 8. Electromyography   175

*Carl Gans*

## 9. Hydrostatic skeletons and muscular hydrostats      205

*William M. Kier*

# Contributors

GARY S. BEAUPRÉ
Rehabilitation Research & Development Center, V. A. Medical Center, 3801 Miranda Avenue, Palo Alto, CA 94304, and Biomechanical Engineering Program, Mechanical Engineering Department, Stanford University, Stanford, CA 94305, USA.

ANDREW A. BIEWENER
Department of Organismal Biology and Anatomy, The University of Chicago, 1025 East 57th Street, Chicago, IL 60637, USA.

R. BLICKHAN
Fachbereich 13 der Universität des Saarlandes, Fachrichtung 4 – Zoologie, D-6600 Saarbrücken, Germany.

DENNIS R. CARTER
Mechanical Engineering Department, Biomechanical Engineering Program, Stanford University, Stanford, CA 94305, and Rehabilitation Research and Development Center, V.A. Medical Center, 3801 Miranda Avenue, Palo Alto, CA 94304, USA.

M. E. DEMONT
Box 159, Department of Biology, St Francis Xavier University, Antigonish, NS B2G 1C0, Canada.

ROBERT DUDLEY
Department of Zoology, University of Texas, Austin, TX 78712, USA.

ROBERT J. FULL
Department of Integrative Biology, University of California, Berkeley, CA 94720, USA.

CARL GANS
Department of Biology, 2127 Natural Science Building, The University of Michigan, Ann Arbor, MI 48109, USA.

J. E. I. HOKKANEN
Department of Medical Physics, University of Helsinki, Siltavuorenpenger 10, SF-00170 Helsinki, Finland.

Contributors

W. KIER
Department of Biology, CB#3280 Coker Hall, University of North Carolina, Chapel Hill, NC 27599-3280, USA.

R. E. SHADWICK
Scripps Institution of Oceanography, Marine Biology Research Division, 0204, La Jolla, CA 92093, USA.

S. M. SWARTZ
Section of Population Biology, Morphology and Genetics, Division of Biology and Medicine, Brown University, Providence, RI 021192, USA.

# Abbreviations

| | |
|---|---|
| A/D | analog to digital (converter) |
| ADB | Apple Desktop Bus (Macintosh) |
| A/P | antero–posterior |
| CCD | charged coupled device (video camera) |
| CM | centre of mass |
| CT | computerized tomography |
| DLT | direct linear transformation |
| DS | downstroke |
| ETOH | ethyl alcohol |
| EMG | electromyogram(s)/electromyography |
| FEA | finite element analysis |
| FEM | finite element method |
| FM | frequency modulation |
| GMA | glycol methacrylate |
| HRP | horseradish peroxidase |
| HYP | hydroxyproline |
| LED | light emitting diode |
| MA | major axis (regression) |
| MEK | methyl ethyl ketone |
| NA | neutral axis |
| OLS | ordinary least squares (regression) |
| RMA | reduced major axis (regression) |
| RS | residual |
| SCSI | small computer systems interface (Macintosh) |
| SI | Système International (d'Unités) |
| SMPTE | Society of Motion Picture and Television Engineers |
| US | upstroke |
| VCR | video cassette recorder |

# 1

# Overview of structural mechanics

ANDREW A. BIEWENER

## 1. Introduction

Organisms must contend with the physical characteristics of their environment as they carry out a wide range of physiological and mechanical functions.
Particular structural solutions must be evolved to meet particular biological
demands, both within and external to the organism. The application of engineering principles to living (and fossil) organisms, and to the materials of which they
are built, represents an important interdisciplinary approach that can facilitate
and help to open new insights into our understanding of organismal design and
function. An understanding of this relationship is also important in terms of how
organisms respond and adapt to changes in their physical environment.
Although biomechanics has seen a recent growth in a number of fields within
the biological sciences, the approach has had a long history (1, 2).

In attempting to understand the mechanical design of organisms, a number
of questions may arise: What are the forces that organisms must generate
and withstand to carry out their mechanical functions? What are the displacements of structural elements resulting from those forces? How are the
material properties of structural elements related to their functional requirements? And finally, how can we assess the relative importance of differing,
but mutually interdependent, mechanical functions to the design of the
organism?

These questions underlie our study of the mechanical function of all living
(and fossil) organisms, encompassing a complex array of solid mechanics,
fluid mechanics, and dynamics. The principles of mechanics outlined here
apply equally well to plants and animals, but chiefly focuses on the mechanical design of animals due to limitations of space and the editor's personal bias.
Whereas the solid mechanics of organisms concern the forces and internal
deformations experienced by *structural elements* of the organism associated
with force transmission, the fluid mechanical function of organisms involves
the physical principles of flow, centring on the mechanism of fluid propulsion
and transmission. Biologically, this is important in the context of aquatic
and aerial locomotion, as well as respiratory, cardiovascular, and feeding
functions. Dynamics constitutes a branch of both solid and fluid mechanics.

For the former, the displacements and accelerations of body parts that are produced by forces acting on them are of interest. For the latter, the movement of fluid within or alongside the organism is the main concern.

The mechanical design and function of organisms occurs at two interdependent levels: materials and structures. This book focuses on the latter, serving as a companion to *Biomechanics—materials: a practical approach* (ed. J. F. Vincent), which discusses methods for studying the mechanical properties of biomaterials. However, it is inescapable that we consider how the mechanical properties of biomaterials influence the design and function of the structures in which they are found. As fairly detailed discussions of fluid mechanics are presented in three of the chapters dealing with the aerodynamics of animal flight (Chapter 5), the design of circulatory systems (Chapter 10), and the hydrodynamics of animal locomotion (Chapter 11), the discussion here concerns mainly the mechanics of solids. Even so, as the treatment is rather brief, the reader may wish to consult standard engineering texts (3, 4) for a more thorough discussion of the topics that are covered.

# 2. Rigid body mechanics

## 2.1 Uniaxial stress and strain

Solid and fluid biomechanics involve the study and measurement of the *forces* acting on materials, structures, and fluids associated with organisms in relation to the resulting *deformations* of those elements, which also can be measured. When external forces are applied to a structure, the structure deforms resisting those forces. At a material level, this deformation involves displacements of the constituent atoms or molecules that are resisted by the interatomic bonds that hold the material together. The ability of the structure to resist these forces depends on its material organization and properties, as well as on the overall organization and shape of the structure itself (see Section 4 below). Given that larger structures can support greater forces and typically undergo greater deformations for a given force, force and displacement are normalized relative to the size of the structure or the amount of material that is being loaded. In solid mechanics, *force* and *displacement* are normalized as *stress* and *strain*. Because a material's ability to resist force depends on the force ($F$) that is transmitted per unit area ($A$) of material, stress ($\sigma$) is defined as,

$$\sigma = F/A \tag{1}$$

acting in the direction of the force perpendicular to the cross-sectional plane of the material in which $A$ is measured. Similarly, deformation or displacement of the material ($\Delta L$) is expressed as a strain by normalizing the change in length to the original 'resting' length (L) of the material:

$$\varepsilon = \Delta L/L. \tag{2}$$

These are often referred to as the *normal* stresses and strains. Stress has SI units of N m$^{-2}$ (or a Pascal, Pa); strain is dimensionless. If a material is subjected to tension, a tensile stress acts in the direction of the applied force and strain is positive, reflecting an increase in the material's length. On the other hand, when a material is loaded in compression, compressive stress is developed and strain is negative, reflecting a decrease in length.

As stress increases within a material, strain also increases. In a linearly elastic, or Hookean, material (which bone or chitin reasonably approximate) increases in stress are proportional to increases in strain and are for the most part time-independent. The graph relating applied stress to the resulting strain is referred to as a stress–strain curve for the material (*Figure 1A*). That is, the rate of force application (within the normal physiological range) does not significantly affect the nature of the material's deformation and can be considered to occur simultaneous to changes in force (in fact, however, nearly all materials show some non-linear elasticity and time-dependent behaviour; see Section 3 below).

## 2.1.1 Yielding and failure

The slope of the initial linear portion of the stress–strain curve defines the elastic modulus ($E = \sigma/\varepsilon$). The elastic modulus is a measure of the material's stiffness in terms of stress and strain. As greater force is applied to a material, increasing stress further, the slope of the stress–strain curve eventually begins to decrease, at which point the material is said to 'yield'. Yielding involves 'plastic flow' within the material as atomic and molecular components of the material are permanently displaced with respect to each other. For rigid materials, like bone and chitin, yielding is typically limited to a small range of deformation, at which point stress falls rapidly to zero as the material fails. The maximal stress reached when failure occurs, defines the failure (or fracture) strength ($\sigma_f$) and failure strain ($\varepsilon_f$) of the material (*Figure 1A*). The stress at which yielding first occurs, though often difficult to discern, correspondingly defines the yield strength ($\sigma_y$) and yield strain ($\varepsilon_y$) of the material.

Most biological materials (particularly less stiff ones, such as elastin, collagen, and cartilage) however, show varying degrees of non-linear elasticity (*Figure 1B*), in which the stiffness (slope of the $\sigma$–$\varepsilon$ curve) of the material changes over differing ranges of stress and strain. In this case, a local tangent modulus (tangent to the curve), or incremental modulus (see Chapter 10), may be used to describe the mechanical behaviour of the material over certain ranges of stress and strain.

For elastic materials, deformations below their yield point are fully recoverable. That is, when the applied stress ($<\sigma_y$) is removed the material returns to its original shape ($\varepsilon = 0$, dashed line *a* in *Figure 2*). Above the yield point, however, not all of the deformation is recoverable due to plastic flow within the material. In this case (dashed line *b*), when the applied stress is removed strain does not return to zero. Instead, the material exhibits

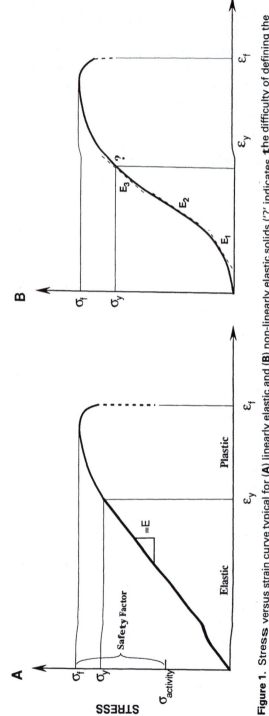

**Figure 1.** Stress versus strain curve typical for (**A**) linearly elastic and (**B**) non-linearly elastic solids ('?' indicates the difficulty of defining the point at which the material begins to yield). See text for the definition and meaning of terms.

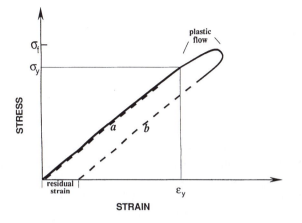

**Figure 2.** Loading (solid) and unloading (dashed) stress–strain curves (**a**) in the elastic range of the material and (**b**) beyond the material's yield strength. When a material is stressed above its yield point ($\sigma_y$, $\varepsilon_y$), plastic flow occurs within the material, resulting in permanent deformation (that is, strain does not return to zero).

residual strain reflecting the extent of plastic flow within the material when loaded past its yield point. For less rigid (low $E$) materials, the amount of yielding prior to fracture can be considerable and may be an important factor in terms of a structure's ability to resist impact loading (see below). For most biological materials, however, mechanical integrity is greatly diminished, if not lost, when substantial yielding occurs. Unlike man-made materials, yield or even fracture damage to biomaterials can often be repaired by the organism, rendering the loss of mechanical function temporary.

### 2.1.2 Safety factor

The ratio of a material's failure stress (or strain) to the magnitude of stress or strain experienced during functional activities ($\sigma_{activity}$) determines a structure's safety factor ($=\sigma_f/\sigma_{activity}$; *Figure 1*). Typically, structural elements of buildings are designed by engineers to have safety factors ranging between 5 and 10. In the same sense, natural selection presumably favours certain safety factors for the structural elements of organisms. Alexander (5) discusses the safety factors of biological structures. Certain factors, such as the cost of building and maintaining a structure, the cost to an organism's fitness when the structure fails, and the level of variability in functional loading patterns, are likely important influences on the particular safety factor that is evolved for a given structural element. In addition, the criterion of failure is important to establishing a safety factor. For instance, if yield failure is the limiting constraint on functional capacity, then a safety factor to yield failure is more appropriate and may be considerably less than if the structure's failure strength were used to establish a safety factor.

### 2.1.3 Shear stress and strain

In addition to experiencing normal stresses and strains, materials also experience shearing stresses and strains. This occurs when a shearing force ($F$) acts in a plane parallel to the material's surface (*Figure 3*). When loaded in this way, the shear stress

$$\tau = F/A_S \tag{3}$$

is defined relative to the surface area ($A_S$) over which the shearing force is applied. The shear strain ($\gamma$) is defined somewhat differently as the angular deformation of the material (measured in radians). The shear modulus ($G$) is defined similar to the elastic modulus as the ratio of shear stress to shear strain,

$$G = \tau/\gamma. \tag{4}$$

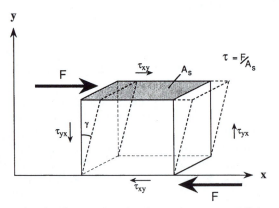

**Figure 3.** Diagram showing how a shearing force, that acts parallel to the surface of a material (having area, $A_s$), produces shear stress ($\tau$) and shear strain ($\gamma$).

### 2.1.4 Elastic strain energy

When a material or structure is stressed, the strain developed within the material involves the absorption of mechanical energy that is stored in the form of elastic strain energy. This strain energy may be dissipated as heat (such as when a wire clothes-hanger that is repeatedly bent back and forth heats up before breaking) or some fraction of it can be recovered (as in the case of a bouncing ball). The amount of strain stored *per unit volume* of material ($U_0$) equals the area under the $\sigma$–$\varepsilon$ curve (*Figure 4A*). For linearly elastic materials,

$$U_0 = \tfrac{1}{2}\sigma\varepsilon. \tag{5}$$

$U_0$ applies to strains in the elastic range of the material and is fully recover-

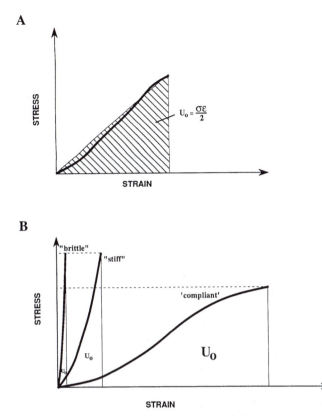

**Figure 4.** (**A**) Stress–strain curve for a 'linearly elastic' material (heavy line), in which the amount of energy stored permit volume of material (area under the curve) can be reasonably approximated by the area of a triangle. (**B**) Stress–strain curves for 'brittle', 'stiff', and 'compliant' materials. Compliant materials (low $E$) store considerably more energy than brittle materials, despite having a lower failure strength (indicated by the dashed lines). Their ability to store large amounts of strain energy makes them difficult to fracture and hence, such materials are considered to be 'tough'.

able when the applied stress is removed. The ability of a material or structure to store strain energy is often the most critical factor determining its ability to resist fracture, particularly under conditions of dynamic loading (that is, when loading occurs at a very high rate). Although a material may be very strong (high failure stress), it will break easily if it cannot absorb much strain energy. Such 'brittle' materials, like glass, deform very little even at very high stresses; in contrast to 'compliant' materials that are capable of much greater elastic deformation (*Figure 4B*). Materials capable of storing large amounts of strain energy are considered to be 'tough'. Toughness is often measured as the work of fracture, or the energy required to break a material.

## 2.2 Principal stresses and strains

When a material is loaded in one direction (uniaxial loading), it also experiences stresses and strains in its opposite dimensions (*Figure 5*). For example, when subjected to tension in one direction ($x$), elongation of the material in that direction causes contraction of the material in the opposing ($y$ and $z$) directions reducing its cross-sectional area in the $xy$ and $xz$ planes. Hence, the material experiences compressive stresses in these latter two directions. Conversely, when loaded in compression along a given axis, shortening of the material along that axis produces extension in its other two dimensions.

### 2.2.1 Poisson's ratio

The relative magnitude of strain in the two planes perpendicular to the plane of applied stress is defined by the Poisson's ratio of the material in those planes,

$$v_{xy} = -\varepsilon_y/\varepsilon_x; \; v_{xz} = -\varepsilon_z/\varepsilon_x. \tag{6}$$

For isotropic materials having uniform properties in all directions, $v_{xy}$ and $v_{xz}$ are equal. Anisotropic materials (such as bone and most other biomaterials), however, show differing amounts of strain in different directions (that is, $v_{xy} = v_{xz}$, and $v_{xy} = v_{yx}$). If the material is incompressible (constant volume), increases in length along the axis of applied tensile stress are compensated for by decreases in cross-sectional area in the opposite dimensions. For isotropic materials, the decrease in $A$ is achieved by equal decreases in the lengths of the other two dimensions (that is, $v_{xy}$, $v_{xz} = 0.5$; or $\varepsilon_y$, $\varepsilon_z = -0.5\,\varepsilon_x$). The applied normal stress and strain, in conjunction with the stresses and strains produced in the opposing planes of the material are commonly referred to as the principal stresses and strains developed within the material. In planar analyses of strain (see Chapter 6), the tensile and compressive strains are defined as the maximum and minimum principal strains, respectively. For conditions of uniaxial loading, the maximum and minimum principal strains are always tensile and compressive, respectively (the larger magnitude of strain defining the axis of applied stress; see *Figure 5*). However, for more complex loading circumstances (that is, biaxial or multiaxial), the maximum and minimum principal strains may be both tensile or both compressive, depending on the orientation and nature of loading.

### 2.2.2 True stress and true strain

Because extension of a material in one dimension causes contraction of the material in its opposing dimensions (or vice versa), the decrease in cross-sectional area ($A$) of the material perpendicular to the applied load also contributes to an increase in stress. In fact, as stress rises within a material, its cross-sectional area changes continuously with the change in stress (decreasing for tensile stress and increasing for compressive stress). This change in $A$

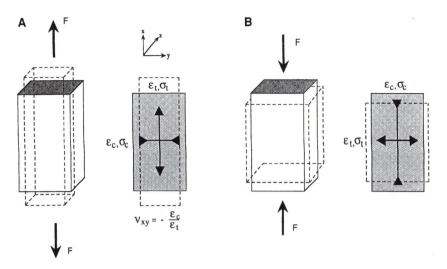

**Figure 5**. Principal stresses and strains developed in response to (**A**) uniaxial tension and (**B**) uniaxial compression, showing the deformations that result for each of these loading conditions.

must be accounted for if the true stress is to be accurately measured. Otherwise, the true stress may vary considerably from the engineering stress calculated using equation (1), in which $A$ is taken to be the unloaded area of the material. For highly extensible materials, such as elastin or spider silk, changes in $A$ are important for accurately determining stress. For rigid materials, such as bone or chitin, the small change in $A$ resulting from increased stress can be safely ignored. Similarly, the progressive change in length of a material as it is loaded also leads to erroneous measurements of true strain if the initial resting length of the material is used to calculate engineering strain. By accounting for the incremental increase in length, true strain is

$$\varepsilon_t = \int_{L_0}^{L} dL/L = \ln L/L_0. \tag{7}$$

## 2.3 Three-dimensional stress and strain

Often, more complex analyses of the stresses and strains developed in three dimensions (3-D) of the material or the structure may be required (*Figure 6*). This involves not only a consideration of the normal strains (or stresses) in each plane of the material, but the shear strains (or stresses) as well. Because stresses and strains developed within a material at a particular location are, in fact, three-dimensional in nature, they are considered *tensor* quantities. As opposed to vector analysis, which involves measurements such as force and velocity in two dimensions, tensors are three-dimensional in nature.

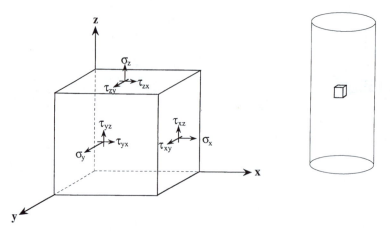

**Figure 6.** Normal and shear stresses (or strains) acting in three dimensions of a cube element of material at some particular location within a structure.

Although a 3-D tensor analysis may provide a more realistic assessment of the complexity of loading experienced by many biological structures, this approach can involve as many as nine material property characteristics to describe a fully anisotropic material (relating normal and shear stresses and strains). In contrast, only two property characteristics ($E$ and $v$, or $G$) are needed to describe a linearly elastic isotropic material. The only feasible way to tackle this kind of problem analytically, as well as computationally, is to employ a more sophisticated, computer-based modelling approach such as that described in Chapter 7. Even so, 3-D stress analyses are fraught with potential difficulties and are time consuming (and hence, costly) to carry out unless the mechanical characterization of the material or structure can be simplified by assumptions of structural symmetry and/or isotropy of the material in two or more of its principal material planes.

## 3. Linearly viscoelastic materials

In contrast to linearly (and non-linearly) elastic materials, in which stress and strain vary in phase, many materials, particularly more compliant ones, show *time-dependent* properties and are referred to as being 'viscoelastic'. Due to viscous interactions in combination with elastic interactions within the material, the magnitude of stress developed in a linearly viscoelastic material depends on the material's rate of loading (or strain rate). In contrast to a linearly elastic material, in which strain energy is fully stored during loading and recovered during unloading and stress and strain vary in phase (*Figure 7*), in a linearly viscoelastic material, energy is dissipated by plastic or viscous flow within the material as the material is stressed. Because of this, stress and

**10**

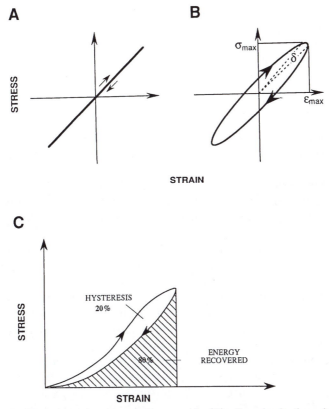

**Figure 7.** Oscillating stress versus strain curves for (**A**) a linearly elastic and (**B**) viscoelastic solid. Whereas stress and strain occur in phase in (**A**), stress and strain occur out of phase in (**B**), making the viscoelastic material properties 'time-dependent'. The extent to which strain lags stress is determined by the phase angle ($\delta$). This typically varies over different loading rates. (**C**) Stress–strain curve for a non-linearly viscoelastic material. Because of its time-dependent properties the unloading curve does not follow the loading curve, even when the material is loaded within its elastic range. This is termed 'hysteresis'. The area enclosed in the hysteresis loop represents the energy dissipated due to viscous interactions within the material, whereas the area under the unloading curve (shaded) is strain energy that can be elastically recovered.

strain vary out of phase with one another (stress leads strain; *Figure 7B*) so that the loading and unloading stress–strain curves show hysteresis (that is, do not overlap) (*Figure 7C*). The area below each curve represents the strain energy stored during loading and subsequently recovered during unloading. The area between the two curves represents the energy dissipated due to viscous flow within the material. Nearly all bio-materials show some viscoelasticity, though for rigid materials like bone, chitin, cellulose and shell this can be reasonably ignored. Collagenous structures, like tendons and ligaments, are viscoelastic but energy dissipation is fairly low (5–10%), so that these structures can serve as good elastic energy storing elements in the limbs of animals (Chapter 3), or as components of the walls of arteries (Chapter 10). Less stiff connective tissues like cartilage and skin, however, exhibit considerable viscoelasticity and their time-dependent mechanical behaviour must be taken into account.

# 4. Composite materials

Most biological structures and indeed, most biomaterials fall within a class of materials termed *composites*. These materials are composed of two or more material components which interact to give the composite material its overall properties. Most often, these materials can be classified as 'fibre composites', having a certain volume fraction of relatively stiff fibres that are embedded in a less stiff matrix. Bone, insect cuticle, wood, and horn (keratin) are some examples of rigid fibre composite biomaterials (in the case of bone, the matrix—calcium phosphate—is more stiff than the fibres—collagen). Cartilage and mesoglea (the connective tissue found in jellyfish and sea anemones) are examples of compliant biomaterials. A discussion of the theory of composites is beyond the scope of this chapter, but given that many of the structural elements considered in this book are built of composite materials, the widespread occurrence and importance of this type of material organization, particularly as it relates to the mechanical properties that are achieved, should be recognized. Much of the time-dependent properties and non-linear behaviour of these materials is due to the interaction of fibres and matrix in transmitting stress through the whole of the material. Readers are encouraged to consult refs. 6 and 7 as an overview to biological composites, and ref. 8 for a more detailed discussion.

# 5. Structures: beam theory

Given that materials are organized to function, either on their own or in combination with other materials, as structural elements within an organism, the stresses and strains developed within them ultimately depend on how they are *structurally* organized in relation to the forces that the structure experiences. For structural elements subjected to simple tension, compression, bending, or torsion the mechanical analysis of structural design is fairly straightforward. However, in most circumstances structural elements of organisms are subjected to a range of loading circumstances that vary over time. In addition, the shape of these elements is often three-dimensionally complex compared to the standard geometries used to simplify the analysis of the effect of shape on stress and strain. Consequently, the engineering analysis must simplify the real biological situation. This is the case, however, even when more sophisticated, computationally intensive approaches are undertaken. For cylindrical elements like arteries and long bone diaphyses, or the cord-like geometry of tendons and ligaments, this simplification is reasonable, providing a valuable analytical approach. In addition, more complex modelling methods still depend on basic mechanics theory applied to standardized structural elements that transmit force and deform in known ways; the complexity of these models stems, in large part, from the number of elements used to model the biological structure.

Although tensile and compressive stresses and strains occur under all modes of loading, it is generally rare that structures experience pure tension or compression. More often, structures are bent or twisted, in combination with being loaded in tension or compression. Unlike tension or compression, in which the cross-sectional shape of the structure is unimportant and stress and strain depend only on the distribution of force per unit cross-sectional area, in bending and torsion the structure's cross-sectional geometry is a very important component of its mechanical integrity. In other words, both the material properties and the shape of a structure determine its ability to resist bending and torsional loads.

Simple beam theory makes three important assumptions in the analysis of stress and strain in relation to structural geometry: that the material is (a) linearly elastic, (b) isotropic, and (c) sections of the material remain planar throughout the range of loads considered. In the discussion that follows the reader once again is encouraged to consult a standard engineering text (3, 4) for a more detailed discussion and derivation of the relationships that follow.

## 5.1 Bending

Consider a cantilevered solid cylindrical beam that is rigidly supported at one end and subjected to a transversely acting force $(-F_t)$ which bends the beam downward (*Figure 8A*). $-F_t$ exerts a bending moment about the beam proportional to $-F_t x$, increasing from the end of the beam to its base ($x = L$) (*Figure 8C*). The effect of the applied bending moment is to bend the beam downward, such that the upper surface becomes convex and the undersurface becomes concave. It is intuitively clear that the upper surface of the beam is in tension (lengthened) and its undersurface is in compression (shortened). Given that the material is linearly elastic ($\sigma \propto \varepsilon$) and isotropic, there must be a linear distribution of stress (or strain) across the face of the beam similar to that shown in *Figure 9A*. Consequently, at some point in the centre of the beam (at the centroid of the beam's section in the $xy$ plane) stress and strain must pass through zero. The axis $(x)$ along which no deformation occurs is defined as the beam's neutral axis (NA). For irregular shapes (or for structural elements composed of materials having different properties), the neutral axis will be asymmetric with respect to the 'upper' and 'lower' surfaces. If the structure is shaped irregularly but constructed of the same material, the neutral axis still passes through the centroid of the beam's section. In bending, then, stress and strain are maximal at the surfaces of the beam (being

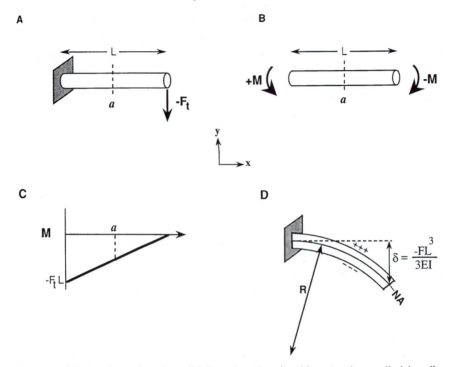

**Figure 8.** (**A**) Cantilever bending. (**B**) Pure bending (in this case the applied bending moment is constant along the beam's length). (**C**) Moment diagram for cantilever bending. (**D**) Deflection diagram of cantilever, showing that the upper surface is subjected to tension (+++) whereas the under surface is compressed (−−−). Near the middle of the beam, therefore, strain (and stress) must pass through zero. The axis along which strain and stress are zero is defined as the neutral axis (NA). $R$ shows the local radius of curvature developed along the neutral axis due to deflection of the beam in bending. $\delta$ is the deflection of beam resulting from the applied bending moment.

tensile and compressive at opposite surfaces; denoted by '+++' and '−−−' signs) and are zero at the neutral axis. A solid cylindrical beam subjected to pure bending is shown in *Figure 8B*; while this is rarely ever encountered by structural elements, it often occurs in combination with axial and transverse loading and is an important component of the stress or strain distribution developed within beam-like elements.

For a beam that develops a radius of curvature $R$ (*Figure 8D*) when subjected to a bending moment ($M$), it can be shown that the distribution of strain (in the y-direction) varies with respect to the distance from the neutral axis as:

$$\varepsilon = y/R \qquad (8)$$

and given $\varepsilon = \sigma/E$,

$$\sigma/y = E/R \qquad (9)$$

**14**

**A  Stress Diagram**

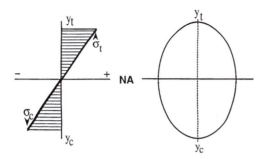

**B  Second Moment of Area & X-Shape**

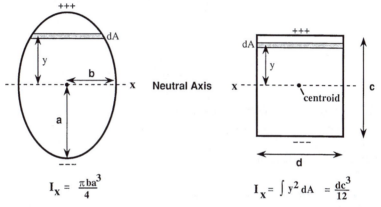

$$I_x = \frac{\pi b a^3}{4} \qquad\qquad I_x = \int y^2 \, dA = \frac{d c^3}{12}$$

**Figure 9. (A)** Diagram of the distribution of stress across a section of the beam that is linearly elastic and isotropic. At right is the beam cross-section (assumed to be elliptical) and the axis ($y$) along which the tensile and compressive stresses act. Stress (and strain) passes from a maximum tensile stress ($\sigma_t$) at $y_t$, through zero at the neutral axis (NA), to a maximum compressive stress ($\sigma_c$) at $y_c$. **(B)** Diagrams of elliptical and rectangular cross-sectional shapes showing how elemental strips of area ($dA$) are summed with respect to the square of their distance ($y$) from the neutral axis to determine the second moment of area ($I_x$) with respect to the neutral axis (defined as the $x$-axis for bending in the $y$ direction). Formulae for $I_x$ are shown below for each of these standard shapes.

This relationship holds along the entire length of the beam. Also, $\sigma/y$ is constant for any section within the beam. Now consider an elemental strip of area ($dA$) at a distance $y$ from the neutral axis but parallel to it (*Figure 9B*). Given that it carries a stress $\sigma$, the force acting on the strip is $\sigma dA$. Because $\sigma/y$ is constant, $\sigma/y = \sigma_t/y_t$ ($=\sigma_c/y_c$) or $\sigma = [\sigma_t/y_t] y$. Consequently, the force acting on the strip ($F_y$) is $[\sigma_t/y_t] y \, dA$ and the moment ($F_y y$) transmitted by the strip $dA$ about the neutral axis is $[\sigma_t/y_t] y^2 \, dA$ ($=[\sigma_c/y_c] y^2 \, dA$). Given that the stresses acting in any section (for example, **a**, *Figure 8*) of the beam act to resist the applied bending moment, the sum of the moments developed by each area element of the section must balance $M$, such that

$$M = \sigma/y \int_{y_c}^{y_t} y^2 \, dA \tag{10}$$

The term $\int y^2 \, dA$ is the *second moment of area,* $\mathbf{I}$ (or area moment of inertia), so that equation [10] can be rewritten as

$$M = \frac{E\mathbf{I}}{R} \tag{11}$$

and by combining equations (9) and (11),

$$\frac{\sigma}{y} = \frac{M}{\mathbf{I}} = \frac{E}{R}$$

or

$$\sigma = \frac{My}{\mathbf{I}}. \tag{12}$$

$\mathbf{I}$ is defined with respect to a given axis; most often the neutral axis. In *Figure 9B*, $\mathbf{I}$ is calculated with respect to the neutral axis ($x$) and hence, is subscripted as $\mathbf{I}_x$. Frequently, $\mathbf{I}$ is subscripted as $\mathbf{I}_{xx}$ to denote it as the *second* moment of inertia about a section.

### 5.1.1 Second moment of area and cross-sectional shape

Because $\mathbf{I}$ depends on the shape of the beam's cross-section, $\mathbf{I}$ will differ along different axes if the beam is asymmetric (elliptical and rectangular sections are shown in *Figure 9B*). For most biological structures, cross-sectional shapes are typically irregular and their geometric properties often cannot be calculated based on formulae for standard geometries. Instead, such shapes must be digitized and their sectional properties calculated using computer algorithms that sum area 'elements' of the section with respect to a defined axis (Chapter 2). Measurements of $\mathbf{I}_x$ or $\mathbf{I}_y$ according to anatomical axes of interest (for example, anteroposterior or mediolateral) may not coincide with the axes of maximum ($\mathbf{I}_{max}$) and minimum ($\mathbf{I}_{min}$) second moment of area values. For this reason, it may be of interest to calculate $\mathbf{I}_x$, $\mathbf{I}_y$, and $\mathbf{I}_{max}$, as well as the angle between $\mathbf{I}_{max}$ and a predefined geometric or anatomical axis. Such measurements, however, still beg the question of the actual direction of bending during functional activities, which for some structures may be determined experimentally (Chapter 6).

$\mathbf{I}$, in effect, represents a measure of the placement of material in a given cross-section with respect to a defined axis. Because each area element ($dA$) is multiplied by the square of its distance $y$, for a fixed area, $\mathbf{I}$ will increase as elements of area are placed at a greater distance from the axis. Further, to minimize the overall deflection of the beam (i.e. keep $R$ small), it is clear from equations (11) and (12) that $EI$ should be maximized and $y$ minimized; or, given that the material has a constant modulus, $I/y$ should be maximized. Stresses developed due to bending will also be smallest when $I/y$ is maximized.

For a given cross-sectional area (and hence, a given volume and weight, if length and density are constant), therefore, a hollow beam has greater 'flexural rigidity' than a solid beam because more of its material is positioned further from the neutral axis. The tubular design of bicycle frames is a familiar example of the need to maximize flexural rigidity and strength per unit weight of material. In engineering construction, the 'I-beam' is frequently used to support vertically oriented bending loads associated with the building's weight. The flanges of the I-beam are displaced from the neutral axis of bending by the intervening web, to give the beam greater $I/y$. The web, like the side-walls of a hollow rectangular or cylindrical beam, is needed to resist shear stresses transmitted along the beam's length (3, 4). Without a shear-resisting element, the advantages of a more distributed shape are lost. In contrast to engineered buildings that can be designed to resist loading in one main direction, biological structures often must support a range of bending loads, in which a more circular or elliptical cross-sectional geometry is desirable. The net moments of inertia for hollow sections can be determined simply by means of subtraction of the values obtained for the hollow region from those obtained for the section as a whole.

## 5.2 Buckling

There is a limit to how expanded the cross-sectional shape of a beam can become, however, before alternative modes of failure are likely. For thin-walled hollow tubes, collapse of the tube's wall in a lateral direction can produce *local buckling* failure, leading to catastrophic collapse of the column or beam, well before general failure due to applied bending occurs. Local buckling generally occurs at a stress $\sigma_L$ given by:

$$\sigma_L = \frac{kEt}{D} \tag{13}$$

where $k$ is an empirically determined constant that generally lies between 0.5 and 0.8, $t$ is wall thickness, and $D$ is the diameter of the cross-section. For a given cross-sectional area, then, $I$ cannot increase indefinitely before the ratio $t/D$ becomes dangerously small.

In long, slender columns, a particular form of local buckling, termed *elastic* or *Euler buckling*, can result due to axial compressive loading. Euler buckling involves a special case of buckling in which a vertical column deflects beyond its elastic stability (that is, local strain and stress exceed $\sigma_y$ and $\varepsilon_y$), leading to local buckling and failure. The magnitude of compressive force that will produce critical buckling is given by the Euler buckling formula:

$$F_E = \frac{n\pi^2\, E\mathbf{I}}{L^2},\qquad(14)$$

where $F_E$ is the axial compressive force to produce Euler buckling, $L$ is the length of the structure, and $n$ is a constant determined by the degrees of freedom of motion of the ends of the column. The values of $n$ can range from less than one (0.25 when one end is fixed and one end is free to rotate in its own plane) to a value of 4 when both ends are firmly fixed (rare in the dynamic world of organisms). The critical shape variables here are $\mathbf{I}$ and length. By expressing $\mathbf{I}$ as $Ar^2$ (where $A$ is the cross-sectional area and $r$ is the least radius of gyration) in equation (14), it is apparent that $F_E$ is proportional to $A/(L/r)^2$, and the resulting failure stress $(F_E/A)$ is inversely proportional to the square of $L/r$, known as the slenderness ratio. The critical stress for very slender columns can be quite low compared to that for axial compression, tension or bending. For bone, the slenderness ratio at which failure is equally likely by Euler buckling versus axial compression is 32 (9) and so, is unlikely to be a critical factor in the design of most long bones.

## 5.3 Torsion

In addition to being subjected to bending, many biological skeletal structures experience torsional or twisting loads. This is particularly so because of the range of mobility required at the joints between articulating support elements. Torsion is analogous to the consideration of bending above, except that the maximal stresses developed are shear stresses, with compressive and tensile normal stresses acting at 45 degrees to the longitudinal axis of the cylinder (*Figure 10*). Shear stress due to torsion is expressed as:

$$\tau = \frac{Tr}{\mathbf{J}}\qquad(15)$$

where $T$ is the torque or torsional moment of the applied force about the structure's neutral axis, $r$ is the perpendicular distance from the neutral axis to the point in the plane of the section at which the shear stress acts (being greatest at the cylinder's surface), and $\mathbf{J}$ is the *polar moment of inertia* (or polar second moment of area) of the cross-section. $\mathbf{J}$ is the sum of the moments of inertia about any two perpendicular axes through the centroid of a cross-section (for example, $\mathbf{J} = \mathbf{I}_{max} + \mathbf{I}_{min}$). Frequently, torsional loads act in combination with bending and axial compressive or tensile loads, making the net loading and consequent distribution of stresses and strains within the beam complex. In general, torsional loading not only promotes shear stress within the structure, but also causes the axes of principal stress (and strain) to deviate considerably from the principal structural axes. The component loads

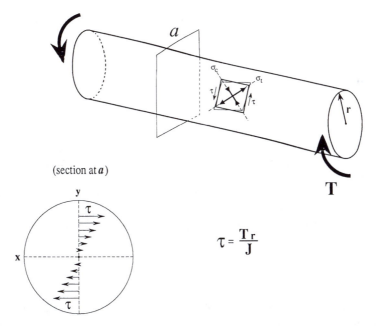

(section at *a*)

$$\tau = \frac{Tr}{J}$$

**Figure 10.** Diagram of torsional loading, showing the resulting orientation of shear and principal stress (and strains for a linearly elastic, isotropic material) in the beam. In this case, torsional stress depends on the applied torque (**T**) in relation to the polar moment of inertia (**J**) of the beam's cross-sectional shape. As in bending, shear stress is zero along the neutral axis, increasing to a maximum value at the beam's surface (along *r*). Shear stresses occur in the plane of the section, whereas principal stresses occur at 45° to the beam's longitudinal axis.

(axial, bending, and torsion) experienced by a structure (that can reasonably be treated as being isotropic and linearly elastic), however, can be determined separately using an analytical approach and combined to determine the net magnitude and distribution of planar stresses and strains. Once again, more involved modelling approaches (Chapter 7) may be required for complex structural geometries and/or anisotropic and non-linear material properties.

## 6. Summary

The non-linear, viscoelastic and/or anisotropic properties of many biological tissues and structures make them annoyingly messy and complex for structural engineers and material scientists who are familiar with engineering materials and structural elements which are designed to be linearly elastic and isotropic. This is made even worse by the ability of many biological tissues to repair themselves, and their tendency to change over time. Because of this, the fairly simple, elementary mechanics presented in this chapter can be usefully applied to identify and understand general, biologically relevant design features of organisms. While more sophisticated mechanical analyses may be carried out, and for certain questions may be appropriate and necessary to achieve useful answers, in many instances the complexity of the real biological situation makes simple mechanical tests of biological function most relevant to the question at hand. In general, it is always wise to begin testing simple ideas and hypotheses, before proceeding to more complicated hypotheses that may entail more difficult experimental tests. It should also be remembered that, as integrative systems, an understanding of the mechanical design and function of organisms must come by studying them at multiple levels of their biological organization. Finally, it is important not to overlook the potential importance of other physiological (i.e. non-mechanical) functions of the systems that we study. For example, although the musculoskeletal system must satisfy mechanical requirements for support and movement of the animal, these must be balanced against the metabolic energy costs associated with carrying them out.

# References

1. Galileo, G. (1974). *Two new sciences*. Transl. S. Trake. University of Wisconsin Press, Madison, WI.
2. Thompson, D'Arcy (1917). *On growth and form*. Cambridge University Press, Cambridge.
3. Beer, F. P. and Johnston, E. R. (1981). *Mechanics of materials*. McGraw-Hill, New York.
4. Faupel, J. H. (1964). *Engineering design*. John Wiley, New York.
5. Alexander, R. McN. (1981). *Sci. Prog. Lond.*, **67**, 119.
6. Vincent, J. F. (1992). *Biomechanics—materials: a practical approach*. Oxford University Press, Oxford.
7. Vincent, J. (1990). *Structural biomaterials* (2nd edn). Princeton University Press, Princeton, NJ.
8. Hollister, G. S. and Thomas, C. (1966). *Fibre reinforced materials*. Elsevier, Amsterdam.
9. Currey, J. D. (1984). *The mechanical adaptations of bones*. Princeton University Press, Princeton, NJ.

# 2

# Shape and scaling

SHARON M. SWARTZ and ANDREW A. BIEWENER

Size influences virtually every facet of an organism's biology, including its biomechanical function. Changes in body size occur during growth (ontogeny), as well as during evolution and may range from tenfold to 100-fold during the growth of an individual to as much as one-thousand-fold to one-million-fold (depending on the breadth of taxa considered) during the evolution of a lineage. Frequently, such large differences in body size require that the shapes of structural components of the organism change to maintain their functional integrity. Analysis of the shape of biological structures, therefore, provides an important empirical approach for identifying and interpreting mechanical constraints underlying the form and function of skeletal support systems. Before discussing practical aspects of methodology in the analysis of shape, we discuss both conceptual and statistical issues related to scaling analyses of shape.

## 1. Effect of size on the mechanical design and function of skeletal systems

### 1.1 Constancy of shape: isometry

The dimensions of a size-series of identically shaped structures will change with size in a predictable way that can be specified a priori on the basis of their geometry alone. When different-sized structures retain the same shape they are considered to scale with isometry or to be 'geometrically similar'. Under these conditions all linear dimensions, such as length ($L$) and diameter ($D$), scale proportional to ($\propto$) body volume ($V$)$^{1/3}$. For structural elements having equal material density, mass can be considered equivalent to volume. Hence, $L$ or $D$ scales $\propto$ body mass ($M$)$^{1/3}$. For example, an eightfold increase in body mass (and volume) corresponds to only a twofold increase in the length and diameter of skeletal elements within animals that are isometric (*Figure 1*). Similar rules can be constructed for other basic shape parameters and functionally relevant variables that can be expressed as dimensional combinations of length, time, and mass (see also ref. 1, Chapters 2–4).

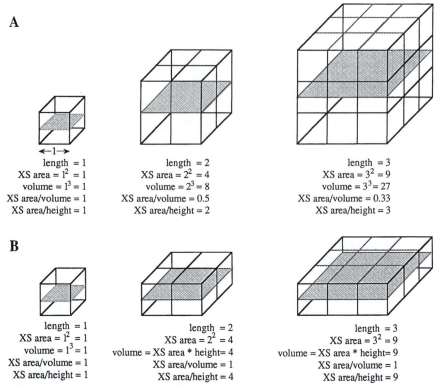

**A**

length = 1
XS area = $1^2$ = 1
volume = $1^3$ = 1
XS area/volume = 1
XS area/height = 1

length = 2
XS area = $2^2$ = 4
volume = $2^3$ = 8
XS area/volume = 0.5
XS area/height = 2

length = 3
XS area = $3^2$ = 9
volume = $3^3$ = 27
XS area/volume = 0.33
XS area/height = 3

**B**

length = 1
XS area = $1^2$ = 1
volume = $1^3$ = 1
XS area/volume = 1
XS area/height = 1

length = 2
XS area = $2^2$ = 4
volume = XS area * height= 4
XS area/volume = 1
XS area/height = 4

length = 3
XS area = $3^2$ = 9
volume = XS area * height= 9
XS area/volume = 1
XS area/height = 9

**Figure 1. A**: A series of cubes illustrating the effects of increasing length on various aspects of size in which shape is constant (isometry). As length doubles, and then triples, the cross-sectional (XS) area of the cube available for supporting weight increases ∝ length squared and volume increases ∝ length cubed. Therefore, cross-sectional area per unit volume decreases dramatically, and the amount of area per unit height increases even though the shape of the three structures is identical. **B**: In this series of solids, cross-sectional area per unit volume remains constant throughout an overall size increase (the same increase in cross-sectional area as illustrated in **A**). In order to maintain this relationship, however, the solids must change shape as they increase in size; this is clear in the comparison of cross-sectional area to height, which goes up much faster than in series **A**.

## 1.2 Mechanical implications of isometry

A basic mechanical requirement of all structures is that stresses be kept within safe limits (see safety factor, Chapter 1). For skeletal elements built of a material having similar mechanical properties, such as vertebrate bone or arthropod cuticle, therefore, peak stresses developed during their use should be nearly uniform if a constant safety factor is to be maintained. This means that *size-related* changes in tissue cross-sectional area available to resist loads must vary in proportion to changes in mechanical loads if stress is to remain constant in different sized organisms. Based on simple geometrical considerations, however, this would seem nearly impossible for organisms to achieve. Given that the forces applied to skeletal elements generally vary in proportion to an organism's weight ($F \propto W^{1.0}$), increases in body weight promote increases in force that exceed increases in the cross-sectional area ($A$, $\propto M^{2/3}$, or $W^{2/3}$ under the constant gravity of Earth). If the shapes of support elements do not change, therefore, stress will be increased with increasing size $\propto W^{1/3}$ (stress = $F/A$), enhancing the risk of mechanical failure in larger organisms. Since functional capabilities depend on the dimensions of structural elements, mechanical function generally must be altered as organisms change size; otherwise, changes in shape must occur.

By analogy, the period of a pendulum as it swings back and forth is solely a function of its length$^{1/2}$. Consequently, if a small and a large pendulum-driven clock are to keep the same time, requiring that their pendula swing at the same frequency, the length of the pendulum of the small clock must be absolutely the same as, or proportionately longer than, that of the larger clock. In this example, a change of shape is required to maintain similar mechanical function. This is not always possible for biological organisms, however, and the need to limit maximal stress (or strain) as size increases may often also involve constraints on mechanical function.

## 1.3 Change of shape: allometry

When one or more aspects of shape change disproportionately with overall size, different-sized organisms are considered to scale *allometrically*, in either a positive or negative way *with respect to* isometry. This patterning can be illustrated in a bivariate scatterplot on a logarithmic scale (*Figure 2*), which shows the relationship between two different, but equivalent dimensions (i.e. a length vs. a length or an area vs. an area). In this case isometry predicts a slope of one, with positive or negative allometry being indicated by slopes either greater than or less than 1. The slope defined by isometry will vary depending on the dimensions of structures that are being compared. For instance, a length (*y*-variable) compared to an area (*x*-variable) will show a slope of 1/2, with allometry being defined relative to this slope. Such allometries may enable certain functional capabilities to remain constant at all sizes. One such model for a positive allometric change in shape is that proposed by McMahon (2) for limb bones of different-sized mammalian species, as well as the branches of trees. In his 'elastic similarity' model, elastic deflections of structural elements are similar at all sizes. For this to occur, larger animals and plants must be progressively more robust, with the length of supportive structures scaling in proportion to diameter$^{2/3}$ (in contrast to isometry where $L \propto D^{1.0}$).

## 1.4 Testing for allometry

By making empirical measurements of the lengths, diameters, and cross-sectional shapes of various structural elements, it is possible to use linear

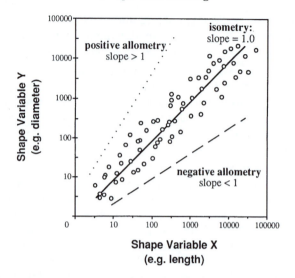

**Figure 2.** A schematic plot of two measurements of equivalent dimension (for example, length vs. diameter shown, or cross-sectional area vs. surface area) on a double-logarithmic scale. After log-transformation, the slope of variables of the same dimensionality will be one if the objects are isometric, and greater or less than one for positive and negative allometry respectively. The null hypothesis about the slope for isometry, however, depends on the dimensions of the parameters of interest. If, for example, the *x*-variable here was volume or mass, the slope for isometry would be 1/3 rather than 1.

regression analysis of logarithmically transformed data to test whether a given series of structures displays isometric versus allometric scaling (Section 2). Isometric scaling can form the basis of a null hypothesis for testing the significance of scaling patterns that deviate from the 'predicted' pattern. Biological data may deviate from the isometric slope either in terms of systematically different regression slopes (indicating either positive or negative allometry) or as systematic deviations of individual data points for particular taxa or individuals from patterns observed for the group as a whole. Deviations of this kind often suggest important functional changes associated with the observed difference in shape.

Setting up alternative hypotheses in scaling analysis requires clear articulation of the predictions that form the basis for each hypothesis, whether they be constancy of shape or constancy of a particular function. Developing hypotheses regarding the relationship between shape and a particular function, in turn, requires a careful determination of which parameters of structural shape are most critical to mechanical function. Chapter 1 introduces the importance of shape in relation to mechanical properties of structures; the considerations discussed there can help to identify suitable variables for allometric analysis.

# 2. Bivariate linear regression analysis

## 2.1 Data sets

Allometric analyses can be carried out on a variety of kinds of data, in particular:

- measurements taken from individuals of a given species or closely similar species at a variety of ages and hence sizes (ontogenetic scaling)
- measurements taken from a size range of adults of a given species (intra-specific static scaling)
- measurements taken from individuals representing a range of species, usually at a comparable, standardized age (interspecific scaling)

Different kinds of biological and biomechanical questions can be addressed by careful sample choice; along with this, it is critical to clearly delineate the hypotheses that are to be tested by allometric analysis so that the data collected are appropriate to answering the specific question of interest (3). In general, it is helpful to sample data from as broad a size range of organisms as possible within the taxonomic range of interest. A broader size sample helps minimize problems arising from relatively 'noisy' biological data, improving the accuracy of statistical estimation of scaling coefficients and exponents (4). Technically, allometric relationships established by statistical methods (see below) are appropriate for predicting values of the two related variables only over the range of the data collected (interpolation). Extrapolation of pre-dicted values outside the range of data sampled is generally unwarranted. Consequently, a broader size range also intrinsically increases the generality of the relationship that is established for the taxa sampled.

## 2.2 Quantifying the scaling relationship: the allometric equation

Delineating the relationship between body size and biomechanical shape variables is fundamentally an empirical, statistical procedure. The scaling relationship between some morphological or physiological variable of in-terest, $Y$, in terms of some basic size variable, $X$, commonly is assumed to take the form:

$$Y = aX^b \tag{1}$$

where $a$ is typically known as the *scaling coefficient* and $b$ is known as the *scaling exponent* (5). In many allometric studies, null hypotheses take the form of specific predictions about the values of these scaling coefficients and exponents. For example, one set of hypotheses may predict that a particular linear measurement ($Y$) will be related to body size ($X$) by an exponent of 0.33 under isometric or geometric scaling, but at an exponent of 0.50 if the function of the structure in question is similar over the size range of interest; similar predictions can be made in some circumstances about the value of the scaling coefficient. A central goal of many allometric analyses is to establish as precisely as possible the values for these parameters and to understand the functional significance about the statistical relationships among these variables.

The allometric equation is often expressed in its logarithmic form:

$$\log Y = \log a + b \log X \tag{2}$$

or,

$$\ln Y = \ln a + b \ln X$$

where the scaling exponent $b$ is now the slope of the line described by this relationship and the log of the scaling coefficient 'log $a$' (or 'ln $a$') is the $y$-intercept. Transformation to either base-10 or natural logarithms is standard practice, with the use of $\log_{10}$ being the most common. In some cases, particularly in growth allometries, there is an a priori expectation that the data should follow a log-linear pattern. However, expressing allometric relationships in logarithmic form is partly a matter of convenience, based on the observation that interspecific morphometric data tend to show log-normal distributions, and because logarithmic transformation tends to equalize the variance in $y$ values over the entire range of $x$ values.

Log-transformation affects the calculated values of scaling coefficients in an important way. Regression lines fit through the raw data will intersect a point representing the mean values of the $X$ and $Y$ variables; the mean of the log-transformed data, however, is the *median* of the original variable, and regressions of log-transformed data will therefore go through the point whose coordinates are [median $(X)$, median $(Y)$] rather than the means. As a result, the antilog of the scaling coefficient determined in this way will be a biased estimate of the scaling coefficient in terms of the original scale of measurement unless the log scaling coefficient is multiplied by a correction factor. The appropriate correction factor is calculated from the standard error of the estimate (SEE) of the scaling coefficient (see ref. 6 for its derivation):

$$CF = \exp\left[\frac{(SEE^2)}{2}\right]. \tag{3}$$

This correction factor is accurate for logarithms taken to base-$e$; for base-10 logs, the SEE must first be converted to base-$e$ by multiplying it by $\log_e 10$ (=2.303).

## 2.3 Regression techniques

A variety of techniques are used to fit regression lines to bivariate data. The appropriateness of different methods has been and currently remains a matter of some debate.

### 2.3.1 Model I regression

The most commonly used bivariate regression technique is the Model I or ordinary least-squares regression (OLS). This is routinely the default (or only) line-fitting technique available in statistical software packages. This method fits a line to bivariate data by minimizing the sum of the squared residuals of the $Y$ term with respect to the line; the residual being the difference between the predicted and measured $Y$ values for each value of $X$.

There are several important, though often overlooked, assumptions underlying this method:

(a) the error term of the regression equation is normally distributed, has a mean of zero and has a constant variance;

(b) for each value of $X$ (termed the independent variable in this method), values of $Y$ (the dependent variable) are normally distributed;

(c) the variance in the dependent variable is constant over all values of the independent variable;

(d) the values of the independent variable are known, or can be determined without error and are set by the investigator.

These assumptions are difficult to meet in most analyses of scaling. In particular, the size variable, be it body mass or any other 'size' measure, is rarely if ever a true independent variable as set out in assumption (d) above. OLS regression of $Y$ on $X$, therefore, will always underestimate the absolute value of the magnitude of the slope of the line. Further, not only are direct measures or alternative estimates of body mass rarely known with less error than the 'dependent' variables (representing dimensional measures of shape), but size values are most often dictated by the choice of sample under study and not set by the investigator.

On the other hand, some investigators feel that body mass holds a central place in allometric analysis, having an underlying causative effect on the variable in question (most often, in this case, a variable of physiological function). In this view, body mass may appropriately be considered as the independent variable determining values of dependent variables. In this kind of circumstance, if the error variance of body mass values is extremely low, it may be appropriate to use OLS regression. It is, however, often preferable, if not necessary, to adopt an alternative regression technique.

### 2.3.2 Model II regression

Model II regression techniques adopt the assumption that both $X$ and $Y$ variables are subject to measurement error as well as biological variability. The two most frequently used techniques of this type are major axis (MA) and reduced major axis (RMA), or geometric mean regression. These methods make assumptions about the relative magnitudes of the error terms of the $X$ and $Y$ variables that differ from OLS regression and from one another. Whereas OLS assumes that the error variance of the $X$ variables is zero, MA assumes that the error variances of the $X$ and $Y$ variables are equal, and RMA assumes that the error variances of the $X$ and $Y$ variables are proportional to the true variances of the $X$ and $Y$ variables. These methods also differ in the way in which the regression line is fitted to the data (*Figure 3*). Whereas OLS regression minimizes the squared distance from each data point to the line in the $Y$-direction only, MA minimizes the squared perpendicular distance from each data point to the line, and RMA minimizes the product of the $X$-direction and $Y$-direction distances from the point to the line.

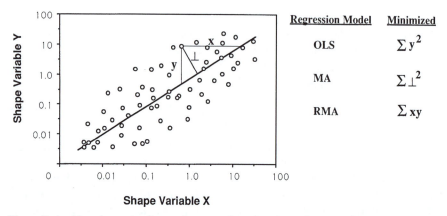

**Figure 3.** In this schematic allometric regression plot, three distances from a given point $(X_n, Y_n)$ are designated: the distance from the point to the regression line in the $y$-direction (**y**), the $x$-direction (**x**) and perpendicular to the line ($\perp$). Different regression methods (shown at right) find the regression line by minimizing the sum of these different distances squared.

In theory, the best way to choose among these alternative models is to know the error variances of the variables. In practice, we often know with certainty that the $X$-variable error variance is large enough to violate the assumptions of OLS, but we may not be able to assess accurately whether MA or RMA regression is most appropriate a priori. RMA has proven to be the most robust method with respect to assumptions about the relative magnitude of error variances and is the least biased estimate of the functional relationship between the two variables in most biological cases (4, 7).

The slope of the RMA regression is the geometric mean of the OLS regression of $X$ on $Y$ and $Y$ on $X$, and is given by:

$$b = \frac{\pm s_y}{s_x} \qquad (4)$$

where $s_y$ and $s_x$ are the standard deviations of the $Y$ and $X$ variables and the sign of the slope is the sign of the sum of cross-products, $S_{xy}$ (8). Alternatively, the slope can be calculated from the slope of the OLS regression as:

$$b_{RMA} = \frac{b_{OLS}}{r_{xy}} \tag{5}$$

where $r_{xy}$ is the correlation coefficient. The intercept, $a$, of the RMA regression is then calculated from:

$$\text{mean (log } Y) = b \text{ mean (log } X) + \log a. \tag{6}$$

Standard errors of RMA regression slopes are approximately equal to those based on OLS regression. These standard errors can thus be used to set confidence limits for the RMA slope, as they would for the OLS slope (3, 9, 10). Because of the ease of computing the slopes, intercepts and statistics obtained using OLS vs. RMA regression, it is generally recommended that both sets of data be included when reporting the results of a scaling analysis, allowing the reader to be his/her own judge.

# 3. Size standards

Analysis of biomechanical parameters with respect to body size requires an appropriate choice of a body size variable. Body weight (or mass) is probably the best possible measure of overall size, but is not without problems. For analyses based on museum collections, and certainly for any analyses of extinct taxa, body mass is usually unavailable. Sometimes mean values for a given species reported in the literature can be substituted, but body weights reported in the literature may be strongly biased by a number of factors (age-distribution of the sample, poor precision of field weight measurements, geographic variation among subspecies, size-biased collection techniques). Geographic variation may be particularly important, as the magnitude of errors due to this effect can easily range from 5 to 40%. Even when body weights are available for the sample under study, complications may remain due to sexual dimorphism of both body size and the structural features of interest and seasonal or nutritional fluctuations of body mass. For pregnant or nursing females (of mammalian species), the 'weight' that the skeleton must support may be 10 to 40% greater than typical body weight.

Alternate estimators of size can be also adopted or may be required when actual body weights are unavailable. Typically these are based on some aspect of the size of the skeleton itself, often a feature that shows strong correlation with body weight within the taxon of interest or a closely related group. Regression equations describing the relationship between the morphological variable(s) and body weight in some reference group of animals are then used to predict body weight when it is unknown (11). Caution must be exercised here, as even the best estimates based on regression equations still may produce rather imprecise values, especially if estimation is extrapolated beyond the size range for which the original regression was calculated (generally an unwarranted practice) or if taxa that differ from the reference group are involved.

# 4. 'Contextual factors' in biological scaling

## 4.1 The phylogenetic context of size and shape comparisons

The phylogenetic background or history of organisms under study is an important but often neglected aspect of the biomechanical analysis of structural design (see ref. 12 for an excellent discussion of the relationship between functional patterns and historical biology). Often, an analysis of size and shape placed within a phylogenetic framework can help to interpret the patterning of structural features in its appropriate functional and evolutionary context. Phylogenetic analysis involves the construction of cladograms to depict the evolutionary relationship of descendent taxa that evolved from earlier ancestors. The methods of cladistic analysis are beyond the scope of this chapter. In brief, a cladogram hypothesizes geneological or evolutionary relationships of a group of taxa based traditionally on morphological characters (traits) that are shared by two or more of the taxa (more recently molecular characters based on genetic sequencing techniques are being used to establish 'molecular phylogenies'). These characters may be as simple as the presence or absence of a particular bone or bony process, or some shape measure, such as the length or area of some skeletal feature, which (central to our interest here) may have biomechanical significance. Based on the character data set obtained by the investigator, computer algorithms can then be used to establish the branching diagram that best describes the relationship of the taxa relative to the 'character state' of a more distantly related taxon.

### 4.1.1 Example of vertebral morphology

A historical approach to scaling analysis can be used in several ways. For example, one might hypothesize that a particular scaling pattern should characterize the size of vertebrae of terrestrial vertebrates relative to body mass if bone form is dictated by the need to resist forces related to gravity acting on body mass. A good fit of empirical data to this prediction would support the hypothesis, but leaves open the question of *why* the data display the particular pattern observed. In other words, although the functional hypothesis may be correct, the possibility that the observed relationship is due to constraints associated with the species' evolutionary history cannot be ruled out. To discriminate between these possibilities requires that a phylogenetic framework be incorporated into the analysis. Framed in a phylogenetic context, because terrestrial vertebrates evolved as a monophyletic lineage from aquatic ancestors (*Figure 4A*), terrestrial vertebrates should not only show a particular scaling pattern but that pattern should also be distinct from that of their non-terrestrial relatives. When the initial hypothesis is supported not only by a good fit of data to the prediction for the group in question but also by clear evidence of the origination of the pattern at the appropriate branching point in a cladogram, the biomechanical hypothesis is much more strongly corroborated.

### 4.1.2 Example of adaptations for digging

Interspecific scaling analyses may also show that a general allometric pattern characterizes some group as a whole, while individual species deviate from the pattern in different ways and to different degrees (often expressed as residuals from a regression analysis). These residuals can be mapped out on the cladogram of the taxa under study in comparison to the pattern of phylogenetic relationship and the distribution of other functionally relevant characters (e.g. presence, absence, or relative development of particular muscles or bones, physiological characteristics, etc.). In this way it may be possible to discern an historical pattern of size and shape evolution, and to observe covariation in scaling features and other functionally important variables (*Figure 4B*). For example, a scaling analysis of bone diameter relative to body mass in mammals might show that, in association with the larger forces of forceful digging, specialized diggers have much greater forelimb bone diameters than one would expect compared to other species of their size. Mapping this characteristic on to a cladogram might indicate several kinds of patterns, including: (a) how many times this biomechanical feature arose in parallel within the mammal lineage; (b) what other musculoskeletal or behavioural features are associated with the evolution of increased forelimb bone diameter; (c) the evolutionary sequence of the acquisition of bone robusticity relative to the other features seen in specialized diggers (*Figure 4B*).

To evaluate evolutionary patterns in structural design, phylogenetic considerations must play a role in the selection of organisms for study. Taxa whose phylogenetic relationships are well-understood are better candidates for scaling analysis than those whose taxonomic affinities are poorly known. Selection of groups where a biomechanical function is widespread within the group but absent in a closely related taxon may aid in the process of clearly distinguishing functionally relevant scaling patterns. And, it may be informative to select cases where similar functional performance is believed to have evolved independently more than once, to allow for multiple independent tests of how size and shape change in particular mechanical contexts.

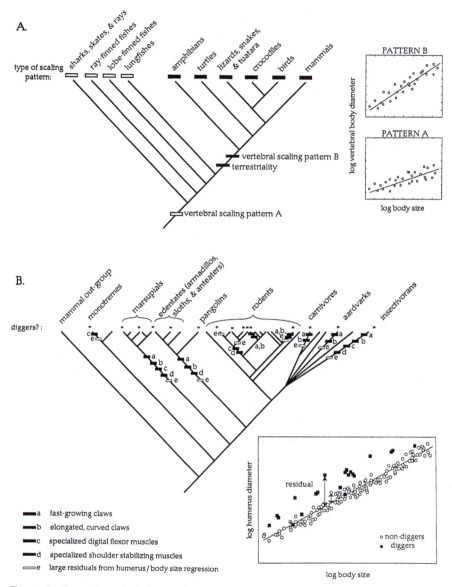

**Figure 4.** Illustration of phylogenetic context for interpreting allometric patterns. **A**: A hypothetical case of changes in vertebral body diameter with changing mechanical environments. If one predicts that the mechanical demands of the evolution of terrestriality should be associated with a particular kind of scaling pattern, then the pattern should (a) be observed within the terrestrial lineage, and (b) be distinct from the pattern of related but non-terrestrial taxa. Here, scaling pattern B characterizes all tetrapod taxa; this pattern appears at the same point on the branching diagram as the first appearance of non-aquatic forms. If, however, the fish taxa displayed scaling pattern B rather than scaling pattern A, it would no longer be justifiable to associate pattern B with the demands of

## 4.2 The developmental context of size and shape comparisons

Size-related changes in form and function have important relationships to developmental processes. Such size-related changes during growth are referred to as ontogenetic scaling. For a given species, growth entails not only enormous changes in size but also considerable shape reorganization. Given the strong dependence of mechanical performance on shape, the mechanical capabilities of a particular structure are also likely to change during growth. Scaling analysis can be used to examine how these ontogenetic changes in shape may influence the organism's development and functional performance during growth.

Developmental considerations are also central to scaling analyses of adults. Either deviations from theoretical predictions based on structural mechanics or conformity to them may arise as a result of basic characteristics of growth processes that may constrain the possible range of adult forms realized. If, for example, there is a critical rate limiting step in the development of a given tissue, the maximum final adult size of structures made of that tissue may be restricted. Similarly, some structures do not grow at all once they are formed (for example, vertebrate teeth). Hence, their postnatal ontogeny cannot show changes in size or shape with changing body size. To determine whether commonalities in scaling patterns among closely related species are due to similar functional requirements or to shared growth trajectories may require reconsideration of adult data in an ontogenetic context (13).

## 5. Measuring shape

### 5.1 Measurements of length

Though simple to make, linear measurements can contain important information for some modes of analysis. Indeed, scaling patterns among and within

---

terrestrial life. **B**: A hypothetical case of multiple evolution of a mechanically specialized behaviour, digging, in several distinct mammalian lineages. Each lineage with digging members is indicated with an asterisk. A variety of conventional musculoskeletal features may distinguish digging animals from their non-digging relatives (characters a–d); scaling analysis of humerus diameter vs. body size also shows that diggers typically have relatively wider humeri than non-digging mammals, and this large positive residual from an allometric analysis can be viewed in the same manner as other morphological characters. When the distribution of characters a–e amongst mammals is studied, it is clear that (i) different digging lineages have solved similar mechanical problems in varying ways; (ii) characters a and b always appear together, and presumably evolved in concert; (iii) the allometric character, e, never appears as the sole digging feature of any group, but (iv) appears independently in several lineages, in association with a wide range of other digging features.

groups of organisms have for the most part been based on the scaling of linear dimensions. Linear dimensions, such as length and diameter, may convey important information about the relative proportions of an organism, and these proportions, in turn, may contain functionally relevant information. Metre sticks, calipers (vernier, dial, and more recently, digital electronic) and micrometres can be used to make straightforward measurements directly from fairly large specimens (their accuracy ranging from 0.1 to 0.01 mm). Digital electronic calipers (from Mitutoyo, Inc., for example) greatly facilitate data collection and storage by enabling direct transmission of the measurements to a personal computer. Smaller specimens or structures can be measured using microscopes equipped with an eyepiece that accepts a calibrated micrometre reticle capable of measuring linear distances in the range of microns. Alternatively, structures of interest may be photographed or traced from microscopic views using a camera lucida with appropriate scale markers. These techniques generally require that the image be positioned appropriately with respect to a length scale in the field of view and carefully focused. Curvatures of skeletal elements, or other characterizations of shape (perimeters, angles, etc.), are generally more easily measured from photographic images than directly from the specimen. By magnifying the specimen, these approaches greatly increase the resolution of the measurements that are taken.

## 5.2 Measurements of cross-sectional shape

The three primary measurements of a structure's cross-sectional geometry that determine its ability to support axial, bending, and torsional loads, respectively, are (a) area ($\mathbf{A}$), (b) second moment of area or area moment of inertia ($\mathbf{I}$), and (c) polar moment of inertia ($\mathbf{J}$). Because many biological structures vary along their length, values of each of these parameters may differ from one location to another location within a structure. Some understanding of the distribution of forces acting on a skeletal member (estimated from models or determined empirically) will help determine where cross-sectional shape is best assessed. In a beam subjected to three-point bending (see Chapter 1), similar to the long bone of a mammal, the bending moment increases linearly from either end to the middle. If the bone's ability to withstand maximal stress (or strain) is of interest, measurement of the bone's area moments of inertia (in two orthogonal axes) at its midshaft is most appropriate. Measurement of cross-sectional shape at other levels, however, may provide insight into how stress is distributed along the bone's length.

### 5.2.1 Computer digitization of shape

Cross-sectional geometry can best be measured from transverse sections of the structure of interest. Working with photographs of sections, digitized coordinate data giving a structure's cross-sectional shape can be obtained from a graphics tablet, or a computer screen image acquired via an image scanner; these data are transmitted to a computer via a serial RS-232 interface or the ADB or SCSI port of a Macintosh computer. In addition to cross-sectional geometry, these devices can be used to obtain linear measurements, including perimeters. A variety of digitizing tablets are available for both IBM-compatible and Apple Macintosh computers that range in size from less than $0.25 \, m^2$ to the size of a large drawing table. These tablets use a cursor with a magnifying cross-hair or a pen-like stylus to digitize coordinates. Cost varies from around US$200 to several thousand for very large, sophisticated table models. Some digitizers are backlit to facilitate clearer imaging, as when working with radiographs, or are semi-opaque for rear projection (GTCO Corp. and Summagraphics, Inc.). The resolution of most digitizing tablets is on the order of $\pm 0.05 \, mm$ (absolute scale). With known magnification of the section (typically by photography or other imaging means) resolution can be correspondingly increased (for instance, $10 \times$ magnification increases digitizer resolution to $\pm 0.005 \, mm$). Images may be projected directly on to the digitizing surface (as for photographic slides or bone sections mounted on microscope slides), traced using camera lucida drawings, or digitized from photographic enlargements. Care must be taken to ensure that magnification of the specimen is known and that both production of the image to be digitized and the digitization routine are accurate and repeatable.

Another means for digitizing and quantifying the shape of structural elements is through the use of video data acquisition systems (see Section 5.3.3 below and Chapter 3). Video cameras can be directly interfaced to computer systems via a 'frame grabber' board for manual digitization via an 'on-screen' cursor (displayed on the video or computer monitor), or automatically digitized based on differential contrast of selected object(s) relative to their background. The quality and nature of lens used on the video camera, however, may cause image distortion. Careful measurements of lengths and areas from good quality graph paper over the entire field of view of the lens (especially near the edges) is one easy way to ensure that this is not a problem (16).

## 5.2.2 Non-invasive imaging techniques—computerized tomography

Computerized tomography (CT) can be used for a non-destructive determination of cross-sectional contours of various anatomical structures in multiple planes and locations (*Figure 5*). Several considerations are critical to obtaining accurate, precise CT scans of skeletal structures. CT scanners collect a variety of radiodensities, coding them black, grey or white, according to user preferences. The image constructed by a CT scanner from the raw data it collects depends on the user's choice of scanner sensitivity. The optimum settings for imaging a particular issue, such as bone, varies from scanner to scanner, from day to day for any given scanner, and according to the particular

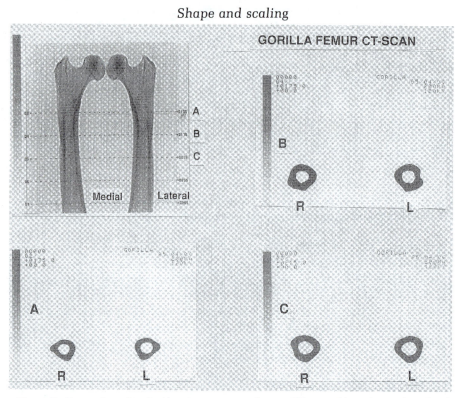

**Figure 5.** Illustration of typical hard-copy output from a CT scan of long bones (proximal gorilla femur). On the left, the bones are viewed from the ventral side, and markers showing the levels from which cross-sectional views will be made are superimposed. Three cross-sectional views are shown at the levels indicated.

size and mineral density of the specimen. It is therefore necessary to calibrate the scanner levels before each series of data collection using, ideally, a sectioned bone of known area that is similar in size to the specimens of interest. It is also helpful to attach reference markers (such as small thin wires or solder) to the surfaces of each specimen to ensure correct anatomical orientation. Finally, care must be taken to orient the specimens within the scanner such that the plane of the image corresponds to the section desired (see also refs. 14 and 15).

## 5.3 Shape analysis software

Commercially available software for shape analysis frequently provide algorithms to determine cross-sectional area, in addition to a variety of other linear measures (see below), but rarely provide the capability for determining the area and polar moments of inertia ($I_{xx}$, $I_{yy}$, $I_{min}$, $I_{max}$, and $J$) of a complex shape. In some cases, large, powerful multifunction software packages will include these calculations (for example, VersaCad and ImageAnalyst for the

Macintosh), but their price may be prohibitive for those simply interested in measuring area and area moments of inertia (US$500–2500).

### 5.3.1 Computer algorithms and sampling considerations

Computer algorithms that calculate $I$ and $J$ from images of complex shapes generally do so by partitioning the structure's shape into discrete area elements, the size of which are limited by the spatial resolution of the digitized data captured from the image. For images that are traced on a digitizing tablet, the speed of data transmission throughput to computer memory (typically up to 100 coordinates per second) and speed of cursor movement are the primary factors affecting spatial resolution of the image. For the case of a scanned or video captured image represented by a pixel distribution of given contrast on a monitor, each pixel represents a discrete area element that can be summed, together with other pixels of similar contrast according to the square of their distance from a defined set of axes (user controlled) to compute $I$ with respect to those axes ($I_{xx}$ and $I_{yy}$, *Figure 6A*).

Traced images from a digitizing tablet, on the other hand, are partitioned into small rectangular shapes, whose individual moments of inertia can be calculated based on standard engineering formulae (17) with respect to the axes of interest and summed in a similar fashion over the whole of the section (*Figure 6B*). Frequently, most rectangular elements summed in this way incorporate either an overestimate or an underestimate of the actual local geometry. This can be corrected for by either subtracting or adding an 'error' value of $I$ based, once again, on a known geometric shape (for example, semi-parabolic arc or parabolic spandrel). In addition, concavities of the structure's contour may incorporate unwanted regions that must be subtracted from the total area before computing the second moment of area of the section. Instead of prescribing fixed axes about which the area moments of inertia are to be calculated, the centroid of the section can be determined and the axes defined with respect to it. In addition to calculating values of $I$ with respect to specified axes that have specific anatomic orientations, the minimum ($I_{min}$) and maximum ($I_{max}$) area moments of inertia, as well as the polar moment of inertia, can easily be computed as well.

### 5.3.2 Customized software

Traced images have been analysed on IBM-compatible computers using the SLICE program (18), or by programs developed and used by individual investigators. A BASIC program, SECTION.BAS (used in refs. 19 and 20) is included on the enclosed diskette for use with SummaSketch Plus digitizing tablet (Summagraphics, Inc.). Another program, MacMoment, is also included for use on a Macintosh computer. Calculation of sectional properties using this application is based on pixel summation of shapes that are input as MacPaint (PNTG) format files created by digitizing tablets, scanners, etc.

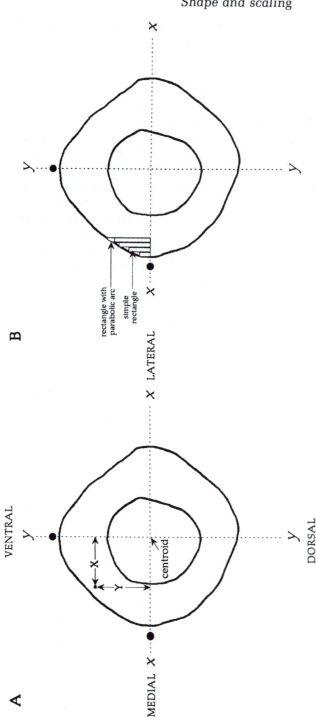

**Figure 6.** Area moments of inertia can be calculated from cross-sectional images in different ways. **A**: With a scanner or video system that inputs a cross-sectional shape to a computer screen, moments of inertia about user-defined axes (here anatomical axes $x$–$x$ and $y$–$y$) are calculated on the basis of the area and position of each pixel that falls within the boundaries of the shape. For $I_{xx}$, pixel area is multiplied by $Y^2$, its perpendicular distance from the $x$–$x$ axis, and for $I_{yy}$, by $X^2$. **B**: With a digitizing tablet, the $x$–$x$ axis is defined by the user (or established from the centroid) and the orientation of the section on the digitizer. Then, as the image is traced with a cursor or stylus, a rectangle is constructed with its long axis perpendicular to $x$–$x$ and its width determined by the distance between consecutive coordinate values on the outlined shape (this is sample rate and cursor speed dependent). The second moment of area of each rectangle with respect to the $x$–$x$ axis is then computed from standard geometric formulae and summed over the entire area. An analogous procedure finds the moment of inertia with respect to the $y$–$y$ axis. Accuracy can be improved by refining the shape of the 'rectangle'; for example, the basic rectangular shape can be replaced by a rectangle 'capped' along the perimeter of the cross-section with one curved side, typically corresponding to a semi-parabolic arc or parabolic spandrel geometry, for which formulae for the second moment of area are known.

### 5.3.3 Other software packages

A variety of commercially available software is also available for simpler geometric data acquisition and analysis using digitizing tablets (such as areas, linear distances, etc.). Two good programs for the Macintosh include Mac-Measure, a public domain application written by the Research Services Branch of the National Institutes of Health, National Technical Information Service (NTIS) that measures areas, perimeter lengths, point-to-point lengths, relative distances, and coordinates of digitized objects, set to a scale and unit of measurement of choice; results can be viewed, printed, or saved to disk for later analysis by other programs. A similar program, Digitize (Rock-ware, Inc.), accomplishes similar results for a low price (around US$150). A useful IBM-compatible application is PC-3D from Jandel Scientific (see Chapter 11 for details); it calculates the same parameters listed above, and can create three-dimensional reconstructions (and thus volume estimates) for a series of two-dimensional images that are a known distance apart.

For images captured by video, there are several kinds of commercial image analysis software on the market. They vary a great deal in speed, flexibility of measurements, statistical capabilities, the degree to which they permit image manipulation and enhancement, etc. There are a number of IBM-based image analysis systems geared for scientific research; one developed with an emphasis toward morphological analysis of organisms is MorphoSys (University of California, Berkeley). This relatively inexpensive software (US$250) is well-suited to acquiring outline data from images of structures of interest, and can easily be used to measure distances, areas, and angles. A more expensive but widely used commercial application is JAVA (Jandel Scientific); it is quite flexible and can easily make linear, perimeter and area measurements. Several Macintosh image analysis programs have also appeared on the market; these include Image (like MacMeasure, a public domain application is available from NTIS), and more expensive image processing alternatives, such as Image Analyst (from Automatix, Inc.) and IPLab (from Signal Analytics Corporation). The more sophisticated video software analysis packages often use edge detection algorithms to reduce the time necessary to capture image outlines; however, care must be exercised to ensure that the process of conversion of the image from the original source through the video camera to the computer screen does not distort the location of shape edges, and this will often require special attention and precise adjustment of image contrast.

## 6. Scaling and shape analysis

### 6.1 Mammalian long bone allometry and brachiating primates

In quadrupedal tetrapods, the forelimbs support the anterior part of the body and exert propulsive forces against the substrate. This generally subjects the limb to compression, and, because long bones are typically curved and because the predominant direction of muscle and reaction forces rarely coincide precisely with the bone's longitudinal axis, the long bones of the limbs generally experience significant bending (21). Compressive and bending loads therefore characterize the biomechanical environment in which the long bones of terrestrial species have evolved. Although this loading pattern is ubiquitous, there are a number of species who constitute important exceptions to the rule. Among these, brachiating apes (gibbons) are unique by having their forelimbs loaded in tension during normal locomotion. In particular, the ulna is loaded almost exclusively in axial tension during brachiation (22). Gibbons rarely use the forelimbs to support their body weight in any way other than suspension. Consequently, their ulnae do not normally encounter significant compressive or bending stresses. Since the ability of a structure to resist tension depends only on cross-sectional area and not on shape (that is, I or P), one would expect a relaxation of selection for bone geometries well-suited to resisting bending.

We have examined this issue by analysing the scaling relationship for the midshaft second moment of area of mammalian limb bones to obtain a background pattern against which to compare brachiating primates. In a geometrically similar series of animals, second moment of area will scale proportional to length$^4$; hence, the expected slope of a log-log regression of I vs. body mass is 1.33. For our mammal sample, the RMA slope is 1.37 (SEE = 0.06), matching the prediction of geometric similarity closely (*Figure 7*). In comparison to other mammals, gibbon humeri appear unexceptional, despite the fact that they may be expected to experience relatively little bending and compression. The gibbon ulna, however, clearly deviates enormously from the general mammalian pattern based on data for femora, tibiae, and humeri. It is plausible and, in fact, an appropriate concern that the small I (and small cross-sectional area) of the gibbon ulna may be a general feature of the ulna compared to the other bones in the sample. However, even if the second moment of area of the gibbon ulna is doubled, the division from the other mammals remains substantial. Clearly, the altered loading regime of the gibbon skeleton is directly reflected in the structural design of the ulna, as seen in its cross-sectional geometry.

## 6.2 Ontogenetic scaling

To date, there remain few biomechanical analyses of ontogenetic allometry. The potential insights of such an approach, however, are well illustrated in an excellent study of the ontogenetic allometry of the limb skeleton and major locomotory muscles in growing jack rabbits (23). In this species, limb bone lengths show significant positive allometries during post-natal growth, whereas most muscle lever arms show significant negative allometries (*Figure 8A* and *B*). As a consequence, young jack rabbits have a *greater* mechanical

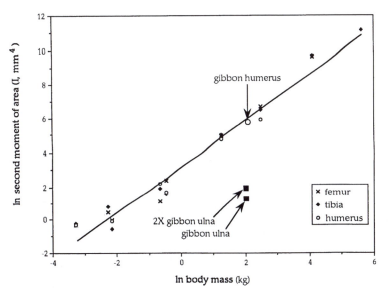

**Figure 7.** Scaling of long bone midshaft second moment of area with respect to body mass in mammals (on logarithmic coordinates). Each point is the mean of several values for one bone of one species, with each represented by a different symbol; the sample includes species ranging in size from mice to horses (data are adapted from (19) and S. M. Swartz, unpublished). There is a regular increase in second moment of area with size ($r =$ 0.97); the slope of the regression is 1.37, close to the value of 1.33 predicted for isometry. In gibbons, a species characterized by a highly derived locomotor pattern that greatly alters forelimb bone loading, the humerus appears no different from typical mammals in cross-sectional geometry, while ulna second moment of area is far less than predicted. This effect cannot be accounted for by the sharing of load between the radius and the ulna, since even if the ulna value is doubled, gibbons deviate substantially from the remaining sample.

advantage (ratio of muscle 'in-lever' to bone 'out-lever') at key limb joints compared to adults, providing them with a superior ability to exert propulsive force for their size. This allometry is evident even when physiological measurements of muscle contractile properties are combined with the simple geometry of the musculoskeletal system: tetanic force generated by the gastrocnemius surprisingly scales in direct proportion to body mass ($M^{1.0}$), rather than $\propto m^{2/3}$, as would be predicted based on the expected change in muscle fibre cross-sectional area (*Figure 8C*). The propulsive force exerted by the foot to accelerate the animal (gastrocnemius force × mechanical advantages $\propto M^{1.0} \times M^{-0.12}$) therefore decreases relative to the mass of the animal ($\propto M^{0.88}$).

The ability of the growing animal's bones to resist forces will be determined both by the material properties of the bone and by its cross-sectional geometry. The mineral content of jack rabbit bones is about 55% early in growth, but

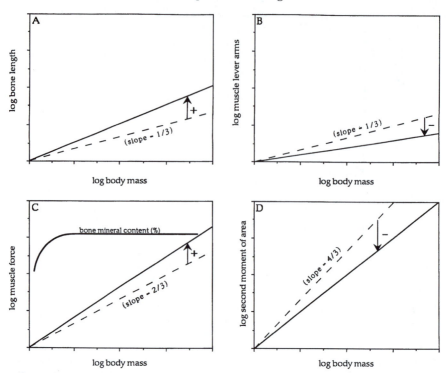

**Figure 8.** Scaling relationships of musculoskeletal features of jack rabbits during post-natal growth. In each schematic, the dashed line gives the allometric slope for geometric similarity and the solid line depicts empirical results. **A**: Bone length scales with positive allometry. **B**: Distances on skeletons representing lever arms for important locomotory muscles scale with negative allometry. **C**: Muscle force scales with positive allometry; the dotted black line shows the concurrent shift from low to normal adult mineral content with age. **D**: Second moment of area scales with negative allometry. Adapted from data in ref. 23.

increases and levels off early in post-natal ontogeny to adult values of about 75%. Bone second moments of area show strong negative allometry, with scaling exponents close to 1.0 during the same period (since the dimensionality of $I$ is $L^4$, an isometric scaling of $I \propto M^{4/3}$ would be expected). This gives young animals a relatively larger $I$ than adults (*Figure 8D*), compensating for the lower mineralization of their bones. As rabbits experience strong selection pressure due to predation, locomotor ability is critical to their survival, particularly at a young age. The shape of the skeleton, both in terms of lever mechanics and bone cross-sectional geometry ($I$), helps to compensate for performance constraints imposed by normal growth processes during that part of ontogeny when the bone material is weakest and the muscles are developing their adult capacity for shortening speed and contractile force.

# References

1. Vogel, S. (1988). *Life's devices: the physical world of animals and plants*. Princeton University Press, Princeton, NJ.
2. McMahon, T. A. (1975). *Am. Natur.*, **109**, 547.
3. Cock, A. G. (1966). *Quart. Rev. Biol.*, **41**, 131.
4. LaBarbera, M. (1989). *Ann. Rev. Ecol. Syst.*, **20**, 97.
5. Huxley, J. S. (1932). *Problems of relative growth*. Methuen, London.
6. Sprugel, D. G. (1983). *Ecology*, **64**, 209.
7. McArdle, B. H. (1988). *Can. J. Zool.*, **66**, 2329.
8. Sokal, R. R. and Rohlf, F. J. (1981). *Biometry: The principles and practice of statistics in biological research* (2nd edn). W. H. Freeman, San Francisco.
9. Clarke, M. R. B. (1980). *Biometrika*, **67**, 441.
10. Leamy, L. and Bradley, D. (1982). *Evolution*, **36**, 1200.
11. Damuth, J. and MacFadden, B. J. (ed.) (1990). *Body size in mammalian paleobiology: estimation and biological implications*. Cambridge University Press, Cambridge.
12. Lauder, G. V. (1990). *Ann. Rev. Ecol. Syst.*, **21**, 317.
13. Shea, B. T. (1985). In *Size and scaling in primate biology* (ed. W. L. Jungers), pp. 175–207. Plenum Press, New York.
14. Ruff, C. B. (1986). *Ybk. Phys. Anthrop.*, **29**, 181.
15. Ruff, C. B. (1989). *Folia primatol.*, **53**, 142.
16. Fink, W. L. (1991). In *Proceedings of the Michigan Morphometrics Workshop* (ed. F. J. Rohlf and F. L. Bookstein). Department of Zoology, University of Michigan, Ann Arbor.
17. Beer, F. P. and Johnstone, E. R., Jr. (1981). *Mechanics of materials*. McGraw-Hill, New York.
18. Nagurka, M. L. and Hayes, W. C. (1980). *J. Biomech.*, **13**, 59.
19. Biewener, A. A. (1982). *J. exp. Biol.*, **98**, 289.
20. Biewener, A. A. and Taylor, C. R. (1986). *J. exp. Biol.*, **123**, 383.
21. Bertram, J. E. A. and Biewener, A. A. (1988). *J. theor. Biol.*, **131**, 75.
22. Swartz, S. M., Bertram, J. E. A., and Biewener, A. A. (1989). *Nature*, **342**, 270,
23. Carrier, D. R. (1983). *J. Zool., Lond.*, **201**, 27.

# 3

# Force platform and kinematic analysis

ANDREW A. BIEWENER and ROBERT J. FULL

## 1. Introduction: forces and motion of rigid bodies

Animals exert forces to move objects in their environment and to move themselves from place to place. Body segment motion, in turn, produces inertial forces that must also be countered by the muscles controlling motion at related joints. Kinetics involves the study of the forces acting on rigid bodies. Kinematics is the study of the motions produced by those forces. Linking the two analyses together, provides a powerful non-invasive approach for studying the biomechanics of animal locomotion, and motor activity in general. For instance, measurements of the forces applied externally to an animal's body (such as ground reaction forces exerted on an animal's limbs when it runs), coupled with kinematic and anatomical data of the limb, allows an analysis of the internal forces that limb muscles must produce to counter externally applied forces. The data obtained from such studies are fundamental to our understanding of how the design of musculoskeletal structures is related to the physical demands placed upon them during various functional activities of the animal in question. In addition, kinematic analysis alone can provide estimates of the inertial loads associated with body segment motion that must be controlled by muscles. Such analyses are particularly important in relation to studies of nervous system control of motor function.

## 2. Design of force platforms

Force platforms are instruments that record the ground reaction forces exerted by an animal when its limb lands on the platform during the 'support phase' of the stride (*Figure 1*). The following criteria are important to the design of a force platform:

- independent measurement of force in three orthogonal planes
- low 'cross-talk' between force components
- high-frequency response
- linear response over a sufficient range and sensitivity of force measurement

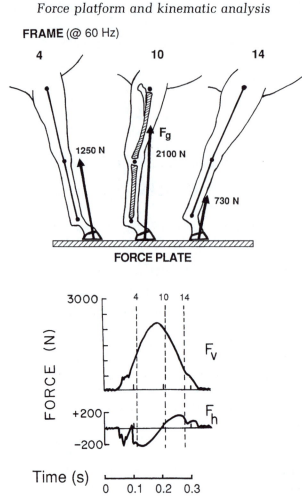

**Figure 1.** Drawings of the position of the forelimb of a horse in relation to the ground reaction force vector ($F_g$) at the beginning, middle, and end of the support phase of a fast trot. Below are the records of the vertical ($F_v$) and horizontal ($F_h$) components of the ground reaction force measured by the force platform. The film of the animal's limb position is synchronized to the ground reaction force recordings by means of a shutter pulse from the camera's shutter, or by some other timing device (e.g. an LED in parallel with a DC voltage source, such as a battery, that is switched on in the field of the video camera can be used to synchronize video tapes to recorded signals).

- uniform response over the platform's surface
- resolution of the point of application or 'centre of pressure' of the ground reaction force
- proper dimensions of the platform's surface in relation to the animal to be studied

Commercially available force platforms (for example, from AMTI & Kistler Instrument Corp.) are designed generally for use in studies of human gait, and hence, are rather large in size. Whereas these platforms are appropriate for use in large animal studies (1–3), they are impractical for studies of small animals. In addition, while the performance of commercially available force plates is good, their cost is comparatively high.

Heglund (4) describes a simple approach for the design of a force platform that emulates the performance of commercially available platforms at a fraction of their cost. The construction and specifications of a force platform outlined here follow Heglund's basic design, but include modifications (introduced by N. C. Heglund and R. Blickhan) that facilitate its construction and enable the measurement of force in three directions.

The ground reaction force ($F_g$) acting on an animal's limb can be resolved into three independent components: vertical ($F_z$), horizontal ($F_x$, fore-aft in direction of the animal's travel) and mediolateral ($F_y$). For species that move their limbs in a para-sagittal plane (for example, cursorial mammals), the mediolateral forces acting on the limb are generally quite small (<5–8%) in relation to the horizontal and vertical ground reaction forces. Consequently, these often can be ignored without incurring significant error in mechanical analyses of the limb. Measurement of only the vertical and horizontal forces also simplifies kinematic analysis of the limb (see below), as films or video recordings need be taken in only a single plane (most commonly in lateral projection). When all three axes are of importance (for example, mediolateral forces are significant in insects and the lateral undulatory locomotion of lizards), three-dimensional analysis of limb movement is required.

Irrespective of whether two or three components of ground reaction force are to be measured, it is essential that the 'cross-talk' between channels be low (typically <3%; see *Protocol 1* below). In addition to minimizing cross-talk, the platform must be designed to have a sufficiently high frequency response (its unloaded natural frequency of vibration) to ensure that the highest frequency components of the ground reaction force that are of interest are faithfully recorded. Generally, the platform's natural frequency should be an order of magnitude greater than the primary signal frequency. In small mammalian species, such as the chipmunk (5), the duration of limb contact at a fast gallop is about 50 msec, corresponding to a primary signal frequency of 10 Hz (stride period: 100 msec). Hence, the natural frequency of the plate should be minimally 100 Hz. To record higher frequency transients in the ground reaction force signal, such as the impact spike present at the start of ground contact of many animals, the frequency response of the force plate must be that much higher.

Obviously, the output of the force platform should be linear in response to applied force in each direction over the full range of forces that are to be measured. Knowing the weight ($W$) of the animal to be studied, the platform can be designed to measure safely the maximally anticipated load ($10 \times W$ is a safe estimate). A trade-off exists, however, between the range and sensitivity of force measurement. In addition to giving a linear response over an appropriate range of forces, the force platform ideally should be insensitive to where on its surface the animal's foot lands. Typically, variation due to the position of applied force should be less than 2–3% worst-case (that is, at the extremes of the plate's surface). This can easily be determined in the vertical direction by placing a known load at varying locations on the plate's surface and comparing the output.

A potentially critical source of error in the analysis of individual limb mechanics is the location of the point of application, or centre of pressure, of the ground reaction force ($F_g$) on the foot. The centre of pressure changes during the support phase, generally progressing anteriorly along the base of the foot to the toes at the end of the support phase. As this can produce large errors in the moments that are calculated to act at joints of the limb, especially when the animal's foot is comparatively long (as in plantigrade and digitigrade species), it is desirable to design the platform with the capability of accurately determining the point of application of $F_g$.

Finally, the overall size of the plate surface depends on the size of the animal (its foot size, stride length, and spacing between contralateral limbs), as well as on the nature of the force analysis to be carried out. In studies of the mechanics of an individual limb, isolated recordings of ground reaction force exerted on one limb must be made. For studies of the mechanical work of terrestrial locomotion (see Chapter 4), the ground reaction forces of all limbs may be recorded through time for one or (preferably) more complete strides. In this case, the force platform must be sufficiently large to span the animal's entire stride length. For larger species, a series of adjacent force platform elements may be required (6, 7).

*Figure 2* shows a schematic drawing of a force platform modified after Heglund's (4) original design. Representative dimensions and performance characteristics of two different sized platforms constructed according to this design are given in *Table 1*. The force transducing elements of the platform consist of two metal beams (front and rear) that support a light-weight, stiff panel which serves as the platform's surface. Honeycomb aluminium panels used in aircraft design (Hexcel Corp.) work well for this purpose; however, cardboard or wood may be used for much smaller animals, such as insects (8). A thin rubber or textured (wall-paper) covering may be applied to a metal panel to provide a non-skid surface for the animal to run over.

The basic design of the platform involves the independent measurement of force in each direction by strain gauges mounted to spring blade elements of the front and rear force beams. Equations of beam-theory are used to calculate the appropriate dimensions of the force beams, according to the mechanical properties of the beam metal, the maximum load limit, and the frequency response of the platform (4). For a maximum applied load ($P$), blade thickness ($h$) can be determined according to

$$h = \left[\frac{6PL}{b\,S}\right]^{0.5} \tag{1}$$

where $L$ is the length of the cantilever formed by the one-half of the blade, $b$

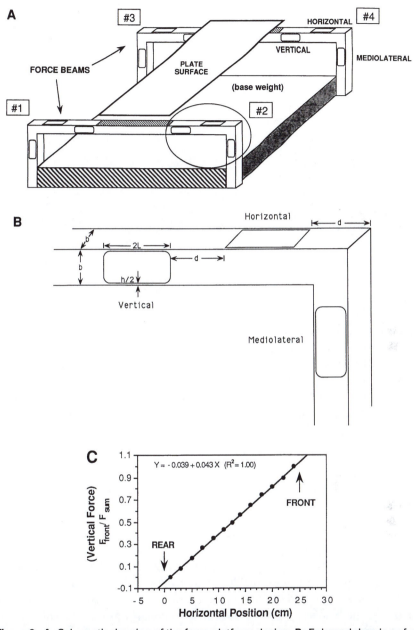

**Figure 2. A**: Schematic drawing of the force platform design. **B**: Enlarged drawing of one set of spring blade elements (vertical, horizontal, and mediolateral) from one beam corner (#2), showing the dimensions which determine the overall sensitivity and performance of the plate, **C**: A representative calibration of horizontal position output based on vertical force ($F_{front}/F_{sum}$), obtained by placing a known weight at specific locations along the length of the plate.

**Table 1.** Specifications for representative force platforms

| | Plate A | Plate B |
|---|---|---|
| Animal weight (N) | 0.5 | 20 |
| Peak load, P (N) | 13.5 | 200 |
| Plate dimensions | | |
|   length (m) | 0.12 | 0.30 |
|   width (m) | 0.06 | 0.20 |
| Surface thickness (mm) | 2.3 | 9.53 |
| material | aluminium honeycomb panel (Hexcel Corp.) | |
| Beam material | Brass | Steel |
|   stiffness, E (GPa) | 105 | 200 |
|   yield strength (MPa) | 105 | 250 |
|   density (kg m$^{-3}$) | 1000 | 7860 |
| Unloaded weight (N) | 0.247 | 13.0 |
| Unloaded natural | | |
|   frequency (Hz) | 400 | 200 |
| Beam dimensions | | |
|   b (mm) | 6.35 | 18.06 |
|   l (mm) | 6.00 | 9.00 |
|   h (mm) | 0.357 | 0.735 |
|   b/h ratio | 17.79 | 25.91 |

is the blade width, and $S$ is the yield strength of the beam metal. The ratio $b/h$ should be at least 14 in order to keep the cross-talk between channels to a few per cent. The weight of the suspended portion of the platform relative to the stiffness of the spring blades determines its natural frequency of vibration. The weight of the beams and the plate surface can be calculated based on their density, once their dimensions are known. The self-loaded natural frequency ($f_{nat}$) of the platform then is

$$f_{nat} = \frac{0.5}{(2D)^{0.5}} \tag{2}$$

where $D$ is the deflection of the plate under these conditions. $D$ can be calculated as

$$D = \frac{4WL^3}{Ebh^3} \tag{3}$$

where $W$ is the suspended weight of the platform and $E$ is the elastic modulus of the beam metal. Because the suspended weight of the beam is greater for more distal spring blade elements (for example, horizontal vs. vertical in *Figure 2*), the blade element furthest from the platform surface will have the lowest frequency response. Accordingly, the force beams should be constructed so that the force component of greatest interest (typically $F_v$) is positioned nearest to the platform's surface.

A BASIC computer program (PLTDSGN.BAS) is included which calculates the design of force beams constructed of a material of known mechanical characteristics (elastic modulus, yield strength, and density) according to these equations.

## 2.1 Strain-gauge and electronic considerations

Either metal foil or semiconductor strain-gauges may be used to transduce force based on deflections of the spring blade elements. The former are less sensitive (lower voltage output per strain), but have greater temperature stability and are easier to work with. An oven-cured epoxy (for example AE-15, Micromeasurements, Inc.) should be used to bond the gauges to the spring blade elements of the metal beam. Heavy duty spring paper clamps work well for clamping the gauges to the beams. The gauges should be clamped during the curing process to ensure a reliable bond. Accurate alignment of the gauges parallel with the blade axis is important for achieving a uniform response from all blade elements. Lead wires from the strain-gauges should be soldered to insulated lead wires via a strain relief connector mounted adjacent to each strain-gauge.

The strain-gauge leads should be configured as a full-bridge (*Figure 3*) for input to a conditioning Wheatstone bridge amplifier (for example, Vishay model 2120, Micromeasurements). Strain gauges of the horizontal and mediolateral force spring elements each should be grouped into a single channel full bridge circuit. To enable determination of the point of application of the ground reaction force in the plane of the animal's motion, however, the vertical force strain-gauges of the front and rear beams should be wired separately as two full-bridge circuits (*Figure 3*), providing independent vertical force measurement at the front and rear of the plate. For uniformity and accuracy, the outputs of the front and rear vertical channels must be the same. This can be achieved by varying the gain of each bridge amplifier channel to give an equivalent output for a load applied at the same relative position to either beam (that is, at the centre of the plate, or at a distance of 10 mm, 20 mm, etc., from either beam).

The voltage outputs of the front vertical, rear vertical, horizontal, and mediolateral channels can be digitally sampled and stored on disk by a computer for subsequent analysis and synchronization with limb coordinate data. Total vertical force is determined by summing the outputs of the front and rear vertical forces in software. The point of force application is then determined by dividing the output of either the front or rear vertical force by the total vertical force. This gives a linear output for position of force application along the length of the plate that is independent of the magnitude of applied force (*Figure 2C*). Resolution of position to within ±0.5 mm is possible.

## 2.2 Force plate calibration

The force plate should be calibrated on a regular basis to check the integrity of the strain-gauges, the electronic circuits and for any changes in sensitivity that may have occurred during its use in previous experiments. *Protocol 1*

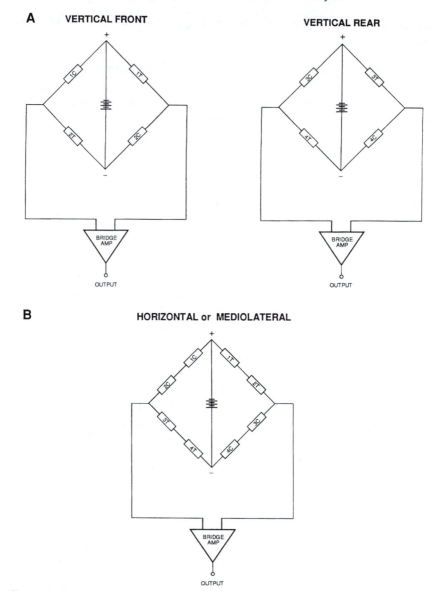

**Figure 3.** Force platform strain gauge bridge configurations. **A**: Vertical force is transduced as independent full bridges at the front and rear beams of the plate to enable determination of the position of force application along the plate's length (position 0 to 25 cm in *Figure 2C*; see text for details). Strain-gauges are numbered corresponding to *Figure 2* (1C, IT, 2C, 2T, etc.). For three-dimensional analyses of the limb, if pairs of strain-gauges (C and T) are configured as four independent half-bridges for each corner of the plate, the centre of pressure of the ground force can be determined at any point on the plate. **B**: Horizontal (fore-aft) and mediolateral forces are each configured as a single full bridge circuit, summing these forces over the entire plate.

outlines a complete calibration and testing of the force plate's performance characteristics. The following supplies are needed:

- metric ruler
- digital voltmeter
- calibrated set of weights (total weight being greater than the maximum expected force that is to be recorded)
- frictionless pulley and cable or force transducer and protractor

---

**Protocol 1.** Calibration procedures

1. Mark off the length and width of the plate at known intervals (say, millimetres or centimetres).

2. Place a weight, equal to or greater than the maximum expected force that is to be recorded, in the centre of the force plate and adjust the gain of the front and rear vertical force amplifier channels to give the same voltage output. Check the uniformity of the front and rear vertical force outputs again at 50% of this weight.

3. To calibrate vertical force, record the front and rear vertical force outputs using a digital voltmeter (the sum of these equals total vertical force, $F_v$) as known increments in weight are applied to the centre of the plate.

4. To calibrate horizontal and mediolateral force, two approaches can be used: (a) a cable (silk suture, nylon cord, etc.) is taped to the plate's surface, passed over a frictionless (air) pulley positioned at the same height as the force plate and known increments of weight are attached to the cable's free end; or (b) the cable is attached to a calibrated force transducer and pulled on by the experimenter, with the angle of force relative to horizontal determined by the protractor. The applied horizontal or mediolateral force is the cosine of the angle times the force measured by the transducer.

5. Sensitivity of the force plate to positional variation in the location of applied force is determined by applying a uniform weight to different locations on the plate's surface and recording the summed output of the front and rear vertical force channels.

6. Cross-talk is determined by recording the horizontal and mediolateral force outputs when a known force is applied in the vertical direction (following this with the same procedure in the other two directions).[a]

7. Position (or the 'centre of pressure') of force application is determined by placing a uniform weight at known intervals along the length (and width, if two-dimensional positional information is needed) of the plate and recording the separate outputs of the front vertical and rear vertical channels using a digital voltmeter. A position calibration curve (*Figure 2C*) is then constructed by dividing either the front or rear vertical force output by the total vertical force.[b]

**Protocol 1.** *Continued*

8. The natural frequency ($f_{nat}$) of the force plate can be determined by lightly striking the platform with a metal object, causing the force platform to 'ring'. The voltage signal of the force platform in each direction displayed on an oscilloscope or computer monitor will show this high-frequency ringing (=$f_{nat}$) superimposed on the primary force signal.

[a] An approximate correction for cross-talk can be achieved by determining the cross-talk between vertical, horizontal, and mediolateral directions at various locations mapped out over the plate's surface. The magnitude of force in each direction due to cross-talk from the opposite two directions can then be subtracted from recorded force signals in software for the various locations on the plate's surface (8).

[b] A convenient way to check the position calibration of the plate is to videotape a ball that is rolled over the plate's length (the weight of the ball will depend on the sensitivity range of the force plate). From these data the centre of pressure is easily determined from the tape.

# 3. Kinematic analysis

Kinematics is the description of the motion of rigid bodies independent of the forces that generate their movement. The following parameters are important for kinematic analysis:

- coordinate system set-up
- image acquisition
- data analysis

## 3.1 Coordinate system

The first step in kinematic analysis is to decide on the motion of interest. If the motion occurs in a single plane, then a two-dimensional coordinate system (for example, $x$, $y$) may be sufficient. If the motion of interest involves two planes, then a three-dimensional coordinate system (for example, $x$, $y$, $z$) must be used. The origin must be located in the appropriate reference frame. If the motion of an animal relative to the ground is important, then the coordinate system should be located with respect to landmarks on the ground. If the motion of an appendage is to be described relative to the movement of the body, then the coordinate system should be located with respect to points on the body (for example, origin at the body's centre of mass). Angles must be defined relative to a zero reference coordinate and should be designated as positive when they change clockwise or counter-clockwise depending on the direction of movement. Fifteen variables can be required to completely describe the movement of a body or body segment [i.e. position of the segment's centre of mass ($x$, $y$, $z$), linear velocity ($x$, $y$, $z$), linear acceleration ($x$, $y$, $z$), angle in two planes ($\theta_{xy}$, $\theta_{yz}$), angular velocity ($\omega_{xy}$, $\omega_{yz}$) and angular acceleration ($\alpha_{xy}$, $\alpha_{yz}$); see Winter (9)].

## 3.2 Image acquisition

Once the coordinate system of the motion has been established, a decision must be made as to the method of image acquisition. A major source of potential error lies in the rate at which images are captured (that is, sampling rate). Sampling theory demands that images must be captured at a minimum of twice the frequency of the motion's highest frequency component. In practice, a tenfold greater sampling rate is usually minimal if frequency and amplitude of motion are to be reliably described. Aliasing errors occur if the sampling frequency is too low (*Figure 4*). These errors can produce a completely false description of motion by generating signals of various frequencies that differ from the frequency of the signal of interest. Higher sampling rates (that is, greater than ten times the motion frequency of interest) are recommended if position data are to be used to determine velocity and acceleration. Small errors in position (phase or amplitude) can lead to enormous errors in velocity and acceleration when position is differentiated directly. Noise is a second reason why sampling rates should not be too low. If the sampling rate is sufficiently high, motion noise can be characterized and later removed (see data analysis). If the frequency of the signal of interest is unknown, then it is preferable to capture the motion at a high frequency initially so that a spectral analysis can be undertaken to determine the frequencies that contribute the most to the signal power. This procedure also helps to identify the frequency range of signal noise, often composed of high frequencies, which can be subsequently removed prior to data analysis. After determining the sampling frequency required, a decision can be made as to which image capture system is most appropriate.

### 3.2.1 Image sampling rate and video cameras

At present, the method of choice for image acquisition is videography (see ref. 10 as a general reference). In the future, the standard is likely to be direct capture of images in computer memory (for example, the Kodak EMS system). A standard video camera (for example, Panasonic model PV series) captures 60 images or 'fields' per second. Each field is composed of 262.5 video scan lines (Standard VHS, NTSC system; 400 lines for Super VHS). Standard playback of a video tape takes two fields and interlaces them to generate a single frame (1 frame = 2 fields). A frame is composed of 525 lines (a composite of two fields captured at 1/60th of second apart) and has a frequency of 30 Hz. Standard frames must be separated into fields to attain a recording frequency of 60 Hz. Video frame grabber computer boards capture a frame and can separate it into two fields. Motion analysis systems (such as from Peak Performance Technologies, Inc.) can use the single fields for analyses.

Higher sampling rates can be attained by video cameras (for example, Kodak EktaPro and NAC-400) that split the fields into multiple pictures.

# Force platform and kinematic analysis

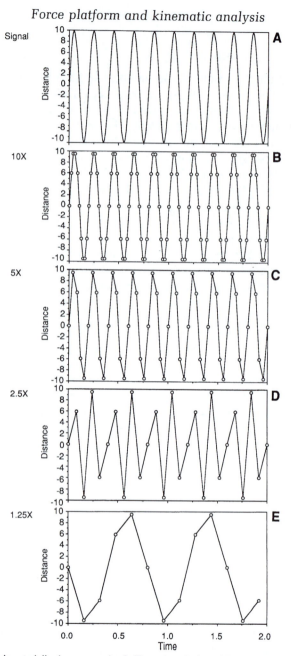

**Figure 4.** Sample rate/aliasing example. **A**: True signal of an object's motion. **B**: Description of motion for sampling rate of 10 times the frequency of the signal; **C**: 5 times the frequency of the signal; **D**: 2.5 times the frequency of the signal; and **E**: 1.25 times the frequency of the signal. Major signal distortion is apparent at 5 times the true frequency and would result in large errors in velocity and acceleration. Lower sampling frequencies completely distort the signal.

Again, it is important to be certain whether the image that you are analysing is a frame, field, or picture (1 frame = 2 fields; 1 field = 1 to 6 pictures). For example, a sampling rate of 120 Hz is attained by splitting a field into two pictures each taken 1/120th of a second apart. As with a standard camera, playback of this image on a standard recorder would show one frame composed of two interlaced fields. Each field would contain not one, but two pictures, usually one below the other. This technique has the advantage of increasing imaging capturing rate, but has the disadvantage of decreasing resolution (usually vertical). Splitting fields in two (120 Hz sampling rate) decreases resolution by 50%, whereas splitting fields into 5 pictures (300 Hz sampling rate) decreases resolution by 80%.

The highest sampling rates (200–6000 pictures/sec) are attained by rapidly moving the video tape during recording (for example, Kodak EktaPro; NAC-400). These systems allow playback directly on standard recorders or downloading to standard systems. However, in some cases a time base correction (that is, sending the signal through a time-base corrector) is required before the field can be captured by the frame grabber board.

To obtain the least blurred image possible it is often desirable to use a strobe light or a mechanical/electronic shutter. Neither system alters the sampling rate, but each can shorten the exposure time of the picture at a given sampling rate, thus helping to 'freeze' the object. Mechanical/electronic shutters on camera lenses are often adjustable so that several different exposure times can be selected (e.g. 1/250–1/10 000 sec).

### 3.2.2 Calibration, lighting, and contrast

Once a camera is selected, its aspect ratio should be determined. Each video camera and lens has a unique ratio of height to width. Video taping a large square of known dimensions is the simplest way to determine this ratio. This correction needs to be made only once for a given camera and lens.

If the motion of interest is confined to one plane, one camera can be used for two-dimensional analysis. The camera should be positioned perpendicular to the plane of motion. Ideally, it should be placed at the height of the subject of interest and levelled. The camera should be placed as far away from the subject as possible to minimize parallax errors. The image should be made as large as possible by increasing the focal length of the lens. This is best accomplished by using a zoom lens with a variable focal length (video cameras either have zoom capabilities or allow a standard C-mount lens to be added). Standard zoom lenses are sufficient for most applications. Lenses that allow greater amounts of light to enter are preferred (that is, low *f*-stops should be used), as high-speed or shuttered cameras typically require more light. A wide open aperture, however, limits the depth of field in which a focused image can be seen; often requiring that a compromise be made between depth of field and exposure time. External lights should be set up to provide the best contrast possible between the subject and the background. In general, the low light levels and heat transmission associated with video recording is a major advantage compared to the considerably greater lighting intensity required for high-speed light cinematography.

After the system is set up, a scale bar of known length should be video taped in the appropriate plane. The scaling factor for the image can be determined from the bar. The bar should fill as much of the screen as possible. A new scale bar should be video taped if the camera is moved or the focal length is changed.

If the motion of interest occurs in two planes, a three-dimensional kinematic analysis is required. A three-dimensional description of motion can be obtained by using mirrors or by using more than one video camera simultaneously. If a mirror is used, it should be positioned at an angle of 45 degrees to the camera's optical axis to give a perpendicular view to that of the camera. The disadvantage of using only one camera and mirrors is the loss in resolution caused by having two or three images contained on a single field. Since the distance of the subject and image in the mirror differ with respect to the camera, focusing may also be a problem. If two cameras are used, they should be placed at approximately right angles to the field of view. A calibration cube or other shape of known dimensions must be video taped in the same location by both cameras. The calibration shape should again fill the screen and must include a minimum of 6 non-coplanar points that can be seen by both cameras (more points are recommended for direct linear transformation; see data analysis, Section 3.3 below). If two or more cameras are used they should be field synchronized (i.e. 'genlocked'). This can be achieved by signalling a common event in the field of view (for example, a light flash ), or by using a SMPTE time code generator that writes the time on the video tape.

In many cases, markers (such as tape, paint, or LEDs) on the subject greatly facilitates digitizing. Markers should provide the greatest contrast possible between the mark and the surrounding background. If the markers are of sufficient contrast, some motion analysis systems have the capacity for auto-digitizing. Remember that markers placed on movable objects, such as clothes or skin, can shift relative to the skeleton as the subject moves.

## 3.3 Data analysis

After the desired image is captured, the motion of interest must be acquired from the video tape. A video taped frame is acquired by a 'frame-grabber' board which resides in the computer (for example, Data Translation, Matrox). A frame-grabber board converts the analogue video signal to a digital representation of the picture. The digital picture is composed of spots of light or 'pixels' that vary in shades from black (0) to white (256). The board has the capacity to separate a frame into two fields. One or more fields can be stored in the board's memory buffer, while the one field is converted back to analogue and sent to a monitor for digitizing. Digitizing is usually accomplished manually by super-imposing a cursor over the picture displayed on the video monitor, locating the point of interest with a mouse, and pressing a mouse button to enter the coordinates into computer memory.

Resolution of the motion of interest is dependent upon the number of pixels that the camera sensor possesses (for example, 286 000 pixels for a standard CCD camera), the number of pixels that are captured and generated by the frame grabber board (for example, 512 × 512), and the number of pixels that can be digitized (for example, 1000 × 1000 ½ pixel precision for Peak Performance Tech., Inc.—the Peak system can average position between adjacent pixels identified by the digitizing cursor, each cursor 'point' being a pair of pixels, thereby achieving a resolution that is twofold greater than the frame grabber board). Pixel numbers ranging from 500 to 1000 yield 0.1–0.2% resolution.

Video controller boards can be used to advance the VCR to the next frame or to find specific frames for continued or additional digitizing. Video controller boards also allow the computer to run the recorder (for example, stop, pause, play, rewind, fast forward). A specific frame can be located by the time code on one of the tape's audio tracks (typically channel 2).

Motion analysis systems (Peak Performance Tech., Inc. and Motion Analysis, Inc.) may possess the capacity for auto-digitizing or tracking. High-contrast markers (such as reflective paint, tape, light-emitting diodes, or dark spots) attached to points of interest on the subject are automatically tracked in software. Tracking is accomplished by first identifying the markers or spots by edge detection (comparison of pixel grey level value with a threshold until an outline is achieved) and then calculating their centres (centroid computation). Prediction of the markers' positions in the subsequent field is done by fitting a function (for example, cubic spline) to the previous coordinate data. Software then performs a search for the next marker or spot position. Data capture is far more rapid than by manual digitizing because frame grabber boards can capture multiple frames for analysis. Auto-tracking can save many hours on large projects. Two problems can occur with this method, however: markers may not be located by the software, or they may 'collide' with one another producing tracking errors. Some motion analysis systems (for example, Peak Performance Tech., Inc.) have error checking routines that identify these problems and allow rapid editing or manual digitizing.

### 3.3.1 2-D analysis

Analysis of movement in two dimensions is accomplished by placing a coordinate system $(x, y)$ onto the field captured by the frame grabber board. The origin of the coordinate system can be digitized as a reference point in every field. This not only corrects for camera motion, alignment, and jitter, but also allows the origin to be placed on a moving object so that relative motion can be determined. Lateral tilt of the camera can be corrected if two reference points $(0,0$ and $x_1,0)$ are used in every field. To increase the speed and ease of manual digitizing, some motion analysis software uses point prediction. The cursor can be placed very near the next point to be digitized by applying a curve-fitting function (for example, cubic spline). Once the position data for a trial have been digitized, the velocity and acceleration of these points can be calculated, but not usually before noise reduction (see data filtering, Section 3.4 below).

### 3.3.3 3-D analysis

Analysis of three-dimensional movement can be accomplished by direct linear transformation (DLT) from data captured on two or more views or cameras (11, 12). The three-dimensional coordinates of a moving object ($x$, $y$, $z$) can be determined from the two-dimensional coordinates ($U$, $V$) that are captured by each camera (*Figure 5*). The position of the object ($x$, $y$, $z$) at any

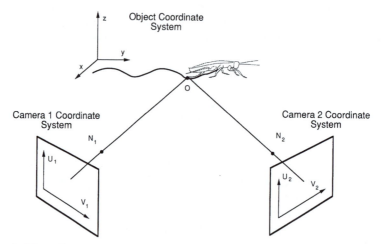

**Figure 5.** Direct linear transformation of two independent sets of two-dimensional co-ordinates ($U_1$, $V_1$ and $U_2$, $V_2$) obtained from two video cameras to three-dimensional coordinates ($x,y,z$). $x$, $y$, $z$ coordinates are related to the camera coordinates by the line $ON_1$.

instant is directly related to a specific set of camera coordinates ($U_1,V_1$ and $U_2,V_2$) by a line or ray that intersects the object (for example, segment $ON_1$; *Figure 5*). The camera coordinates are related to the object coordinates by the expressions:

$$U = \frac{Ax + By + Cz + D}{Ix + Jy + Kz + 1}$$

$$V = \frac{Ex + Fy + Gz + H}{Ix + Jy + Kz + 1}$$

(4) and (5)

where coefficients $A$ through $K$ represent constants which depend on the location of the cameras relative to the object. The values for the 11 coefficients can be determined if 6 or more non-coplanar control or calibration points are measured initially. Once the values for the constants are known, digitized points from the two-dimensional camera coordinates can be used to calculate the three-dimensional object coordinates $(x, y, z)$. There are no specific limitations on camera placement when using this method which makes it convenient due to constraints frequently encountered in many experimental set-ups. However, error increases as the angle(s) between the cameras deviate from 90 degrees. Also, it is important to note that lens distortion may increase error significantly when determining 3-D kinematics. Corrections for various type of lens distortion (for example, linear, barrel, pin-cushion) can be incorporated in the DLT as additional coefficients.

## 3.4 Noise reduction

Raw kinematic data always contain noise due to camera vibrations, video-tape imperfections and digitizing inaccuracies. These errors may be small if only position data are required, but can be enormous if the position data are used to determine velocity and acceleration. This results from the fact that differentiated noise increases linearly with frequency. Several techniques have been used to reduce noise. These include:

- Chebyshev least-squares polynomial curve fitting of displacement data followed by polynomial differentiation (13)
- second-order finite difference (14)
- spline function curve-fitting
- digital filtering followed by finite difference techniques (15)

A variety of noise-reduction techniques should be tried by investigators on their particular data sets. Previous comparisons of the filtering techniques have been undertaken. Zernicke *et al.* (16) found that orthogonal polynomial fits tend to over-smooth data compared to cubic spline fits. Pezzack *et al.* (17) found similar results for polynomials up to the 16th degree. Second-order finite difference differentiation tends not to smooth the data enough and results in noisy acceleration data (17). Digital filtering followed by first-order finite difference techniques often provides satisfactory signals (15, 17).

A fourth order, zero-lag Butterworth digital filter is a fast, general purpose filter (9). The filter allows selective rejection of frequencies above or below a given frequency (this is the cut-off frequency, $f_c$). In most cases, a low-pass filter which attenuates high frequencies is used, because noise is usually of high frequency. The filter sums weighted adjacent raw data and previously filtered data according to the following equation:

$$F_{(i)} = a_1 F_{(i-2)} + a_2 F_{(i-1)} + a_3 R_{(i)} + a_4 R_{(i+1)} + a_5 R_{(i+2)} \qquad (6)$$

where $F_{(i)}$ is a filtered value, $R_{(i)}$ is a raw value, $a_x$ are filter coefficients, and $i$ is the current picture number. The coefficients are determined from the ratio of the sampling frequency to the cut-off frequency.

Filtering can introduce a phase shift in the filtered data. This can be eliminated, however, by filtering the data in the forward direction ($i = 1$ to the total number of pictures) and then in the reverse direction (last picture to $i = 1$). By adopting this procedure, use of the above equation (6) produces a fourth order, zero-lag digital filter.

Selection of an appropriate cut-off frequency is important. The signal near the cut-off frequency is always slightly attenuated depending on the 'steepness' of the filter (that is, how sharp is the filter cut-off). Therefore, if the cut-off frequency is too low with respect to the signal frequency range of interest, signal distortion will occur. On the other hand, if the cut-off frequency is too high, too much noise may pass through the filter. Residual analysis provides an objective method to determine an appropriate cut-off frequency (9). By using a computer, a raw signal can be filtered at a range of cut-off frequencies. The residual ($RS$) at a given cut-off frequency can be calculated by:

$$RS(f_c) = \left[\frac{1}{N}\sum_{i=1}^{N}(Y_{ri} - Y_{fi})^2\right]^{1/2} \tag{7}$$

where $Y_{ri}$ is the $i$th raw data point, $Y_{fi}$ is the filtered data point and $N$ is the sample size. If $RS$ is plotted as a function of $f_c$, the degree of noise versus signal distortion can be evaluated (*Figure 6*). The area above the segment **ae** and below the horizontal intercept line represents the noise residual. The intercept of the segment (**de**, extended to **a**) is the residual mean square error and is the mean noise for the raw data set. If the sampled data contain no

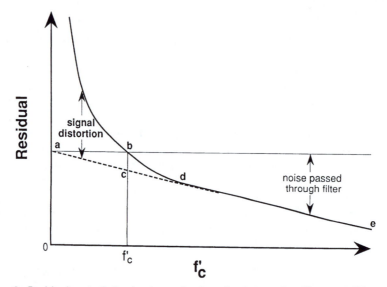

**Figure 6.** Residual analysis for the determination of an appropriate filter cut-off frequency. Frequency $f'_c$ is the frequency that balances the amount of noise versus signal distortion. See text for details.

signal, just random noise, the residual plot would be a straight line decreasing from an intercept on the ordinate at 0 Hz to an intercept on the abscissa at the Nyquist frequency (0.5 times the sampling frequency). Signal distortion is shown by the steep rise in the residual as the cut-off frequency is decreased (below $f'_c$). From this graph, one can select the amount of noise and signal distortion that is acceptable. A reasonable selection would be to balance the amount of noise and signal distortion. The noise and signal distortion become equal at the cut-off frequency where the horizontal intercept line intersects the residual function (that is, point **b**). The equality or balance is shown by the fact that segment **bc** represents both the residual noise and signal distortion.

## 3.5 Velocity calculations

Linear and angular velocities can be determined by using a fourth-order central difference calculation:

$$V_{(i)} = (F_{(i-2)} - 8 F_{(i-1)} + 8 F_{(i+1)} - F_{(i+2)})/(12 \Delta t) \tag{8}$$

where $V_{(i)}$ is a velocity value, $F_{(i)}$ is a filtered position value, $i$ is the current picture number and $\Delta t$ is the time difference between pictures. The advantage of this calculation over the simple calculation of average velocity is an improved precision that is achieved by determining the instantaneous velocity at picture $i$, using the position values both before ($i - 1$, $i - 2$) and after ($i + 1$, $i + 2$). Acceleration can be determined in a similar manner by using a first- and second-order central difference (14).

# 4. Integrating force and kinematic data: linked segment, free-body analysis

By combining ground reaction force data obtained from a force plate with joint coordinate data obtained from a video analysis system, a free-body analysis (*Figure 7*) can be carried out to calculate joint moments, muscle forces, and bone and tendon stresses (or strains; see also Chapter 6). The limb is treated as a series of rigid segments that are linked together by frictionless pivots (the joints) having from 1 to 3 degrees of freedom of movement (for a planar analysis, only 1 degree of freedom of motion is assumed). For each joint of interest, the moment exerted at the joint at a given instant in time must be calculated. The joint moment is the sum of three components: (i) external moment, (ii) inertial moment, and (iii) gravitational moment. The external moment produced by the ground reaction force is simply

$$M_{\text{ext}} = F_{\text{g}} \times R \tag{9}$$

where $R$ is the orthogonal distance from the vector of ground force to the rotational axis of the joint. The rotational axis of a joint is best determined by

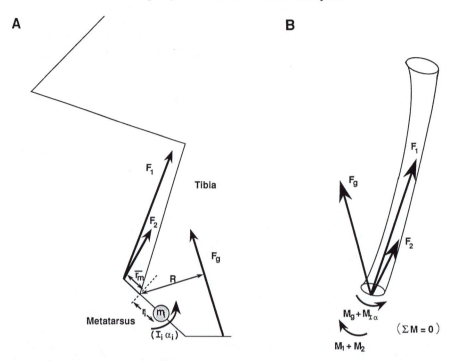

**Figure 7.** **A**: Schematic drawing of a limb showing the external forces that act on the tibia at a specific time during the support phase of the stride ($F_g$: ground reaction force; $F_1$ and $F_2$: muscle forces). In addition to the moment exerted by the ground reaction force ($M_{ext}$), the muscles must also counteract the moment of the metatarsus due to its inertia ($I_i$) and angular acceleration ($\alpha_i$). During most of the support phase, the inertial moment ($M_{inert}$) is small compared to $M_{ext}$. **B**: Free body diagram of the tibia showing the external forces acting on the tibia. Equal and opposite forces (not shown) act to compress and bend the bone, satisfying the conditions of static equilibrium ($\Sigma F = 0$ and $\Sigma M = 0$).

taking a series of radiographs with the joint held in different positions (three is usually sufficient for motion in a single plane) over its functional range of motion. For a three-dimensional analysis, radiographs must be taken with the joint rotated in a second plane (non-coplanar). The intersection of the axes of articulating segments at each joint angle identifies the joint's centre of rotation. This can also be done at the limb surface but is less satisfactory. Markers should be positioned on the overlying skin. The external moment can be decomposed into its component vectors as $M_x = F_x \cdot x$; $M_y = F_y \cdot y$; and $M_z = F_z \cdot z$, where $F_x$, $F_y$, and $F_z$ are the components of the ground reaction force and $x$, $y$, and $z$ are the respective distances of these components to the joint's centre of rotation.

The inertial moment generated due to the motion of limb segments is

$$M_{inert} = \sum_{i=1}^{n} I_i + m_i r_i^2 \, \alpha_i \tag{10}$$

where $n$ is the number of segments distal to the joint of interest, $I$ is the moment of inertia of each segment about its own centre of gravity axis, $m$ is the mass of each segment, $r$ is the distance of the centre of gravity of each segment to the joint's centre of rotation, and $\alpha$ is the angular acceleration ($\text{rad s}^{-2}$) of each joint segment. These are summed for all segments distal to the joint, with the rotational acceleration of each distal segment measured *relative* to the proximal articulating joint segment (see Chapter 4, Section 3, and refs. 9 and 18 for further details). Gravitational moments due to the weight of each segment are determined as

$$M_{\text{grav}} = \sum_{i=1}^{n} m_i\, g\, x_i \tag{11}$$

where $g$ is gravitational acceleration ($9.81\,\text{m s}^{-2}$) and $x_i$ is the horizontal distance of the centre of mass of each segment to the joint's centre of rotation. The net total joint moment then is

$$M_{\text{tot}} = M_{\text{ext}} + M_{\text{inert}} + M_{\text{grav}}. \tag{12}$$

During the support phase of an animal's stride, the moments produced by segment inertia particularly at more distal joints, as well as those due to gravity, are typically quite small in relation to those exerted by the ground reaction force and hence, can often be ignored with little error incurred. However, if the animal's limb segments are a large fraction of its total body mass, as in humans, errors incorporated by ignoring segment inertia can be significant, being increasingly large at more proximal joints (19). Schneider and Zernicke (20) give a detailed mathematical description of intersegmental dynamics for a human limb and a FORTRAN subroutine that can be adopted for calculating the component joint moments in the local plane of the articulating segments.

The magnitude of muscle force required to counter the joint's moment is determined by satisfying the condition of static rotational equilibrium (in which the net moment = 0) at a given instant in time

$$F_{\text{muscle}} \cdot \bar{r}_{\text{m}} + M_{\text{tot}} = 0$$

or

$$F_{\text{muscle}} = -M_{\text{tot}}/\bar{r}_{\text{m}} \tag{13}$$

where $\bar{r}_{\text{m}}$ is the weighted mean moment arm of the muscles that are active to counter the moment at the joint. For relatively simple joints, in which only a single muscle is active to counter the joint moment or all muscles have nearly the same line of action and moment arm at the joint (such as the ankle or elbow joint of mammals), $\bar{r}_{\text{m}}$ is easy to determine. It is simply the orthogonal distance from the muscle's line of action to the joint's centre of rotation, which can be determined from dissection and radiographic measurements (see above). These joints are considered to be statically determinant. However, many joints (e.g. knee, hip, or shoulder of mammals) that are controlled by muscles possessing differing lines of action and moment arms are statically indeterminant. For these joints, no exact solution exists for how muscle force is partitioned between agonist muscles, making it difficult to estimate the moment arm or mechanical advantage of the muscles as a group. In some cases direct *in vivo* recordings of muscle forces (see Chapter 6) can provide an experimental means for quantifying the contribution of individual muscles to total agonist muscle force.

Various optimization criteria and methods can be used to solve for the distribution of muscle force through time among individual muscles acting about an otherwise indeterminant joint (see ref. 21, for example). These approaches use either some quantification of the intensity of muscle electrical activity (EMG; see Chapter 8) or calculate the distribution of muscle forces based on one or more criteria, such as minimizing muscle power output or peak muscle stress. Recently, computational methods based on optimal control strategy have also been developed to tackle this problem (22). In this case, iterative solutions are obtained for the distribution of muscle forces that maximize a certain global criterion (such as maximum jump height). Following Alexander (23), we (24) have made the assumption that the force and hence, the distributed weighting of a muscle's moment arm, will vary in proportion to the fibre cross-sectional area of the muscle. This approach assumes that muscles develop equal stress, minimizing the peak average stress developed within any one muscle of an agonist group. However, given that agonist muscle groups are typically composed of a heterogeneous distribution of muscle fibre types, the recruitment of motor units for muscle force generation often may be unevenly distributed among individual muscle agonists, so that this assumption must be treated cautiously.

Despite the assumptions (and limitations) inherent in the calculation of muscle forces using a non-invasive linked segment, free-body approach, changes in muscle force development can be usefully related to changes in muscle fibre length to estimate muscle power output (but see Chapter 4), as well as being used to calculate tendon strain energy storage and bone stress.

## 4.1 Muscle–tendon length changes

Changes in overall muscle–tendon length ($dL_{tot}$) are calculated from changes of joint angle ($d\theta$), in which $dL_{tot} = r_m \, d\theta$, for small $d\theta$. $d\theta$ may occur in two different planes, requiring a 3-D analysis of segmental motion. However, as noted above, motion of a limb or of a joint in many instances occurs within a single plane. In these cases, a planar analysis of joint motion is sufficient, greatly simplifying the complexity of the analysis involved. The remaining discussion considers a planar analysis of segment motion, in which the joint coordinate data are obtained from a lateral projection of the animal's limb.

To determine muscle length changes requires that changes in tendon length ('series elasticity') be subtracted from overall muscle–tendon length change. Knowledge of the changes in tendon length, in turn, requires that tendon strain ($\varepsilon_{tend}$) be determined (see below). Calculation of actual muscle fibre length changes within the muscle requires that the fibre architecture of the muscle is known.

## 4.2 Tendon strain, tendon strain energy, and muscle length change

Tendon strain and strain energy storage can be calculated knowing tendon elastic modulus ($E$), tendon cross-sectional area ($A$) and tendon volume ($V$). Tendon volume is most easily calculated from tendon mass, if the density of the tendon is known. Tendon density can be determined by Archimedes' principle (18), in which the piece of tendon is first weighed in air (as normally determined on an analytical balance) and subsequently re-weighed while immersed in a fluid (water) of known density. The density of the piece of tendon is given by

$$\rho_{tend} = \rho' \left[ 1 + \frac{W'}{W - W'} \right] \tag{14}$$

where $\rho'$ is the density of the fluid, $W$ is the weight of the tendon in air, and $W'$ is the weight of the tendon while being hung fully immersed in the fluid. This method works quite well for objects (such as tendon) that are of similar density to water because errors in weight do not produce the same errors in density; rather, the error is with respect to the *difference* in density between the object and the fluid. A value of $1120 \, \text{kg m}^{-3}$ has been reported for vertebrate tendon (25). Mean tendon cross-sectional area, then, is: $A = V/L$, where $L$ is the length of the external tendon that is excised and weighed (it is critical that the tendon be weighed immediately to avoid desiccation; small tendons must be weighed to the nearest $0.1 \, \text{mg}$). Alternatively, the cross-sectional area of larger tendons can be determined at specific sites along the tendon's length by cutting transversely through the tendon with a sharp scalpel and photographing the cut end of the tendon with a calibrated length measure in the camera's field. The photographs can then be magnified and digitized to determine local changes in tendon area (see Chapter 2). Once tendon area·is determined, tendon stress is calculated as

$$\sigma_{tend} = F_{muscle}/A \tag{15}$$

yielding

$$\varepsilon_{tend} = \sigma_{tend}/E. \tag{16}$$

Tendon strain energy is calculated assuming that the average cross-sectional area of connective tissue contained *within* the muscle equals the mean $A$ for the cut piece of tendon (that is, equal stress is assumed to be transmitted by the collagen within the muscle to the collagen in the external tendon). This is done by treating the entire length of the tendon ($L_{tend}$) as passing from the muscle's point of origin to the tendon's insertion distally; so that tendon strain energy is

$$U_{tend} = 0.5 \cdot \sigma_{tend} \cdot \varepsilon_{tend} \cdot A \cdot L_{tend} \tag{17}$$

where $A \cdot L_{tend}$ is the total tendon volume strained by the force of the muscle. Finally, the length change of the muscle is

$$dL_{muscle} = dL_{tot} - (\varepsilon_{tend} \cdot L). \tag{18}$$

## 4.3 Muscle fibre area

The measurement of muscle fibre cross-sectional area is straightforward for parallel fibred muscles but less certain for pinnate muscles. Muscle volume is determined from muscle mass, based on a value for the density of muscle (vertebrate striated muscle generally has a density of $1060 \, kg \, m^{-3}$; ref. 18). If there is reason to suspect a different density for the muscle of interest, the muscle's density can most easily be determined by Archimedes' principle (see above). For a parallel fibred muscle, muscle fibre area ($A_{mus}$) is simply muscle volume divided by the muscle's mean fibre length. For most parallel fibred muscles, fibre length does not vary much. However, in cases where muscle fibre length varies considerably throughout the muscle, carefully distributed measurements of muscle fibre length should be made to determine a mean fibre length. The use of a dissecting microscopic with a calibrated reticle in the eyepiece is especially useful for small muscles with short fibres (in the range of 2 to 20 mm). Otherwise, dial metric calipers can be used.

It should be noted that measurements of muscle fibre length are more accurately measurements of muscle *fascicle* length. At present, there is growing debate concerning the actual length of individual fibres within a muscle (see Chapter 8, Section 3). For many parallel fibred muscles, what appears to be a single 'fibre' running end to end within the muscle, actually may be several interdigitated fibres that taper and overlap in series with one another. This raises important questions with respect to the neural activation (fibre recruitment) of the muscle and details of force transmission within the muscle (between the actively contracting fibres and the connective tissue that binds them together). However, for the purposes of estimating a muscle's overall force generating potential or the average stress developed within the muscle, the issue of serially interdigitated shorter fibres is not a serious concern with respect to determining muscle fibre cross-sectional area.

For pinnately fibred muscles, the determination of muscle fibre area is complicated by the variable number of planes along which the fibres are angled, the variable length of the fibres within a plane and the variable angle of pinnation of the fibres. Typically, fibres change their angle of insertion on to the muscle's central tendon or sheath as one moves from the proximal to more distal locations in the muscle. Consequently, a number of measurements of fibre length and pinnation angle distributed throughout the whole of the muscle are required to accurately sample the muscle. It is best to devise an objective scheme for sampling the muscle to achieve unbiased estimates of mean fibre length and pinnation angle. Knowledge of the muscle's internal architecture is also essential if it is to be sectioned properly for making measurements of fibre architecture.

*Protocol 2* can be followed for making measurements of fibre geometry in both pinnate and parallel fibred muscles.

---

**Protocol 2.** Measurement of muscle fibre cross-sectional area

1. After being dissected free, the muscle should be carefully cleaned and weighed. At this point, muscle density can also be determined if necessary.

2. The muscle may be placed in a weak fixative (10% formalin) over night to improve the integrity of the tissue. However, formalin fixation causes shrinkage. This can be checked by comparing measurements of fibre length to those obtained from fresh tissue.

3. Using a sharp scalpel, the muscle should be cut transversely along its length *parallel* to the plane of an angled sheet of fibres within the muscle. If the muscle is multipinnate, then additional sections of the muscle must be made. This should be planned carefully, as making more than two or three sections often renders orientation of the muscle difficult. For small muscles, a dissecting microscope should be used.

4. Support the muscle so that the cut surface is level and facing upward. Cotton soaked in saline works well for this purpose. Ensure that the muscle approximates its 'resting' shape.

5. Sample and record muscle fibre length at standard intervals throughout the plane of section. Again, a dissecting microscope with a calibrated reticle is useful for very short fibres; however, in most instances a metric digital caliper works best.

6. Using a clear, lightweight protractor supported just over the surface of the muscle and a clear plastic ruler, measure angles of pinnation at the same locations where fibre length was determined. This measurement requires that the axis of the tendon of insertion (assumed to be the muscle's line of force transmission) be defined.[a]

7. After making measurements of muscle fibre length and angle on the fresh or mildly fixed muscle, muscle fibre length may be measured a second time, following acid digestion of the connective tissue to verify the original set of measurements. This procedure, however, precludes measurement of fibre angle. To carry out this procedure, immerse the muscle in 15% nitric

**Protocol 2.** *Continued*

acid and check after periods of 8, 16, and 24 hours, until individual muscle fibres can gently be freed from surrounding connective tissue using watch-maker's forceps. The lengths of individual muscle fibres can then be easily measured. Torn fibres are readily discerned from intact fibres by checking the ends under a dissecting microscope: torn fibres show abrupt tears or jagged ends in contrast to the slow taper of intact fibres.

[a] An alternative approach is to photograph the surface of the muscle and make the measurements of fibre length and fibre angle on the photographs. Though more time-consuming and costly, this step typically yields more accurate measurements. The values obtained from the photographed sections can be compared to the direct measurements to determine the necessity of this step.

The fibre area of a pinnate muscle can be calculated by the following equation given by Calow and Alexander (26):

$$A_{mus} = \frac{m \sin (2\alpha)}{2 \rho_{mus} \, l \sin (\alpha)} \tag{19}$$

where $m$ is the muscle's wet mass, $\rho_{mus}$ is the muscle's density, $l$ is the muscle's mean fibre length, and $\alpha$ is the mean pinnation angle. See ref. 26 for details concerning how the muscle's fibre geometry is modelled to derive the above equation.

## 4.4 Bone stress

Bone stresses are calculated by summing the net transverse ($F_T$), axial ($F_A$), and mediolateral ($F_L$) forces acting on the bone in question. In most cases, a planar analysis of bone stress is carried out in which either antero-posterior (A/P) or mediolateral forces are ignored, and no assessment of torsional loading or eccentric loading of the bone out of the plane of motion can be made. More commonly forces acting in the A/P plane are of interest (particularly for cursorial species that move their limbs in a para-sagittal plane) and mediolateral forces are ignored. Mediolateral forces must be considered, however, if a three-dimensional analysis of bone stress is required. The above components of force are generated by the combined effect of joint reaction forces transmitted by the ground reaction force, by muscular forces and by ligamentous forces, if present (*Figure 8*). The transverse and axial force component of each force vector are resolved in relation to the longitudinal axis of the bone (for example, for muscle force: $F_T$muscle $= F_{muscle} \cdot \sin \theta$; $F_A$muscle $= \cdot \cos \theta$). Axial compressive stress at the bone's midshaft then is

$$\sigma_A = -F_A/A_{bone} \tag{20}$$

where $A_{bone}$ is the midshaft cortical cross-sectional area of the bone. Stress due to bending is

$$\sigma_B = \pm(F_T \cdot l + F_A \cdot r_{curv}) \, c/I \tag{21}$$

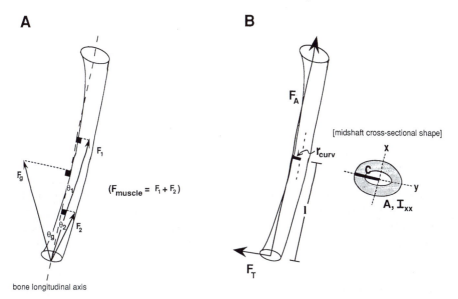

**Figure 8. A:** Free body diagram of the tibia, showing the axial and transverse components of each force vector relative to the bone's longitudinal axis (dashed lines). Axial components of force are determined by $F \cdot \cos \theta$ and transvere components by $F \cdot \sin \theta$. **B:** Net axial ($F_A$) and transverse ($F_T$) forces acting on the bone. In addition to the bending moment produced by $F_T$ at the bone's midshaft ($= F_T \times l$), due to the bone's longitudinal curvature $F_A$ also exerts a bending moment ($F_A \times r_{curv}$). Stresses due to axial compression ($\sigma_c$) and bending ($\sigma_b$) at the bone's midshaft are determined based on geometric properties of the bone's cortex at its midshaft (note that the calculations shown involve a planar analysis of the limb, limiting the computation of stresses to the antero-posterior plane, or the $y$-axis, of the bone).

where $l$ is the distance from the articular surface of the bone to its midshaft, $r_{curv}$ is the moment arm of the axial component of force acting about the bone's own longitudinal curvature in the plane of interest, $c$ is the maximum distance from the neutral axis to the surface of the bone and $I$ is the second moment of area of the bone at its midshaft (measured in the antero-posterior plane; see Chapters 1 and 2). For straight bones ($r_{curv} = 0$), no moment is exerted by $F_A$. The orientation of bending due to $F_T$ and the bone's own curvature determines the distribution of stresses across the bone's cortex in the antero-posterior plane. Although $r_{curv}$ is typically small in magnitude, the presence of longitudinal curvature in many long bones greatly affects the distribution of stresses developed across the bone's midshaft cortex (1, 5). Consequently, it is critical that it be measured accurately. $r_{curv}$ in the A/P plane is best determined from a lateral radiograph of the bone, in which the longitudinal axis of the bone is established by the centres of rotation of the proximal and distal articular ends of the bone. $r_{curv}$ is the distance from the longitudinal (chord) axis of the bone to the centroid of the bone's cross-section. For fairly symmetric bones, the centroid may be assumed to be midway along the bone's diameter, making the measurement of $r_{curv}$ more straightforward. Bone curvature and $F_T$ are additive if $F_T$ acts in the same direction as the concavity of the bone's curvature. For the case in which these are in the posterior direction of the bone's shaft, the stresses at the bone's anterior and posterior midshaft surfaces are respectively:

$$\sigma_{ant} = \sigma_A + \sigma_B; \quad \sigma_{post} = \sigma_A - \sigma_B \qquad (22, 23)$$

$\sigma_{post}$ will be compressive (negative), and $\sigma_{ant}$ will be tensile (positive) if $\sigma_B$ exceeds $\sigma_A$.

This method provides a reasonably accurate means for determining the maximum stresses developed *generally* within a bone in a single plane. It is sensitive however to errors in measuring the magnitude of $F_T(l)$, which depends on the accuracy with which bone orientation to the ground reaction force and muscle vectors can be determined. Although the error may be small in terms of the absolute magnitude of $F_T$, because of a large value for $l$ in long bones this can have a large effect on the calculation of stress due to bending at the bone's midshaft. In addition, this method provides little detail as to the distribution of stress or strain within the bone's shaft, which is exacerbated by ignoring out-of-plane forces. For this, *in vivo* measurements of surface bone strain (see Chapter 6) or finite element modelling methods (see Chapter 7) are necessary. *In vivo* measurements of bone strain can, in turn, be used to calculate or to verify the net components of axial and transverse force acting on the bone, based on an empirical value for the elastic modulus of bone which is needed to transform recorded strains to stresses (1).

The two important advantages of a combined force–kinematic analysis are that (i) it is non-invasive, minimizing concerns about the performance of the animal being studied, and (ii) it integrates estimates of bone stress and tendon stress (or strain) with the underlying muscle forces and ground reaction forces that must be generated to support and move the animal. Finally, this approach is crucial to establishing realistic values for the functional loads experienced by a bone element that are required for input to finite element models of the bone (see Chapter 7).

# References

1. Biewener, A., Thomason, J. J., Goodship, A. E., and Lanyon, L. E. (1983). *J. Biomech.*, **16**, 565.
2. Kram, R. and Powell, A. (1989). *J. appl. Physiol.*, **67**, 1692.
3. Cavanagh, P. R. and LaFortune, M. A. (1980). *J. Biomech.*, **13**, 397.
4. Heglund, N. A. (1981). *J. exp. Biol.*, **93**, 333.
5. Biewener, A. A. (1983). *J. exp. Biol.*, **103**, 131.
6. Cavagna, G. A., Heglund, N. C., and Taylor, C. R. (1977). *Am. J. Physiol.*, **233**, R243.

7. Heglund, N. C., Cavagna, G. A., and Taylor, C. R. (1982). *J. exp. Biol.,* **79,** 41.
8. Full, R. F. and Tu, M. (1990). *J. exp. Biol.,* **148,** 129.
9. Winter, D. A. (1990). *Biomechanics and motor control of human movement.* John Wiley, New York.
10. Inoue, S. (1986). *Video microscopy,* p. 584. Plenum Press, New York.
11. Abdel-Aziz, Y. I. and Karara, H. M. (1971). Direct linear transformation from comparator coordinates into object space coordinates in close-range photogrammetry. American Society for Photogrammetry. Falls Church, VA.
12. Walton, J. S. (1981). *Proc. Int. Soc. Optical. Eng.,* **291,** 196.
13. Kuo, S. S. (1965). *Numerical methods and computers,* p. 234. Addison-Wesley, Reading, Massachusetts.
14. Miller, D. I. and Nelson, R. C. (1973). *Biomechanics of sport,* p. 246. Lea & Febiger, Philadelphia.
15. Winter, D. A., Sidwall, H. G., and Hobson, D. A. (1974). *J. Biomech.,* **7,** 157.
16. Zernicke, R. F., Caldwell, G., and Roberts, E. M. (1976). *Res. Q.,* **47(1),** 9.
17. Pezzack, J. C., Norman, R. W., and Winter, D. A. (1977). *J. Biomech.,* **10,** 377.
18. Alexander, R. McN. (1983). *Animal mechanics* (2nd edn). Blackwell Scientific Publications, London.
19. Wells, R. P. (1981). *Bull. Prosthet. Res.,* **18,** 15.
20. Schneider, K. and Zernicke, R. F. (1990). *Adv. Eng. Software,* **12,** 123.
21. Herzog, W. *J. Biomech.* (In press).
22. Pandy, M. G. and Zajac, F. E. (1991). *J. Biomech.,* **24,** 1.
23. Alexander, R. McN. (1974). *J. Zool., Lond.,* **173,** 549.
24. Biewener, A. A. (1983). *J. exp. Biol.,* **103,** 131.
25. Ker, R. F. (1981). *J. exp. Biol.,* **93,** 283.
26. Calow, L. J. and Alexander, R. McN. (1973). *J. Zool., Lond.,* **171,** 293.

┌─────┐
│  **4**  │
└─────┘

# Mechanical work in terrestrial locomotion

R. BLICKHAN and ROBERT J. FULL

## 1. Introduction: average versus oscillatory work

The amount of metabolic energy required to travel a given distance by an animal using legged, terrestrial locomotion exceeds the costs of swimming by a factor of more than eight and of flying by more than a factor of four (1). Since terrestrial locomotion is expensive, it is reasonable to assume that the relatively high metabolic power input should generate a considerable, and therefore easily measurable, amount of mechanical power output. Surprisingly, a simple and direct link between metabolic and mechanical power during terrestrial locomotion has proved to be elusive (2). Fortunately, the challenge of investigating the link has been rewarding because these studies have and will continue to increase our knowledge of musculoskeletal function for all modes of locomotion.

During walking or running on the horizontal, a body normally returns to the same total energy level once per stride (a stride being one complete cycle of leg movements). If an animal is moving straight ahead at a constant average speed, then the average kinetic ($\bar{E}_{\text{kin}}$) and potential energy ($\bar{E}_{\text{pot}}$) remain constant from stride to stride. The *average mechanical energy* of the system, the sum of potential and kinetic energy per stride, is therefore unchanged (assuming drag is negligible). However, by this measure, efficiency defined as mechanical power output divided by metabolic energy input, would be zero because no total mechanical work is done. When walking uphill, muscular work is necessary to increase the total potential energy from step to step and efficiency is positive (3). Yet, according to this definition, walking or running downhill decreases total potential energy and results in a negative efficiency. Obviously, a consideration of average mechanical energy alone is insufficient to explain how muscles generate movement on land.

The zero or negative work paradox can be explained partially by considering fluctuations or *oscillations of mechanical energy* within each stride as opposed to average energy changes (4). During a single stride, a body's centre of mass accelerates and decelerates in at least the horizontal direction, as well as changes potential energy by rising and falling. These changes in potential and kinetic energy increase the instantaneous mechanical energy. Transient increases in mechanical energy during a stride can be generated from muscular work and, therefore, should be considered in estimates of the mechanical power of terrestrial locomotion.

Cavagna and others (5, 6, 7) have estimated mechanical energy output from fluctuations in the energy of a human's or animal's centre of mass. By using a force platform, they determined the extent to which energy of the centre of mass changed due to the interaction with the environment (that is, due to a leg pushing against the ground and the ground exerting a force on the centre of mass). The mechanical work done by a body on the environment is termed *external work*. External work has been used with respect to the lifting and acceleration of a body's centre of mass during a stride, but more commonly refers to the lifting of weights, the pedalling of a bicycle ergometer or the work used to lift a body's centre of mass uphill.

Analysing the dynamics of the centre of mass has been invaluable in developing general models of terrestrial locomotion. However, as Winter (8) points out, the external work of the centre of mass may not represent the sum of all the energy changes in each segment of the body. He suggests that *internal work*, changes in the energy of segments relative to the centre of mass, should be summed with external work to estimate mechanical energy output. Even the general method of summing segment energies tends to underestimate energy generation and absorption at different joints (9). The calculation of *joint and muscle work or muscle power* takes into account energy generation, absorption and transfer at each joint and can provide a better estimate of muscle work (10).

In the sections that follow, we show the methods used to calculate external and internal energy oscillations as well as joint and muscle work or power. In the remaining part of the chapter, we evaluate these methods and discuss how energy exchange and transfer, elastic strain energy storage, co-contraction of antagonistic muscles and the cost of isometric and energy absorbing muscular contractions can significantly affect the link between musculoskeletal function and the mechanical energy output of terrestrial locomotion.

## 2. Measurement of external work

During straight ahead locomotion at constant, average speed, the fluctuations of external ground reaction forces result in instantaneous work done on the body's centre of mass (CM) and generate corresponding fluctuations of total, external mechanical energy. The external forces generated by the musculoskeletal system to accelerate and decelerate the body's CM can be measured using force platforms or force platform tracks (ref. 11; and see Chapter 3.2). The velocities and displacements of the CM can be obtained by successive integration of the forces. These can be used in turn to calculate the changes in kinetic and potential energy of the CM.

## 2.1 Calculation of energy fluctuations of the CM from ground reaction force measurements

For an animal of known weight, the three components of the acceleration of the CM can be obtained by dividing the corresponding forces by the animal's mass.

The horizontal ($_1$) and lateral ($_2$) force ($F$) yield acceleration such that:

$$a_1 = F_1/m; \ a_2 = F_2/m \tag{1}$$

where $m$ is mass of the animal and $a$ is horizontal or lateral acceleration of the CM. The vertical ($_3$) component of the CM acceleration equals:

$$a_3 = (F_3 - mg)/m \tag{2}$$

where $g = 9.81 \ \mathrm{m\,s^{-2}}$. Weight can be obtained by averaging the vertical force for an integral number of strides. This calculated weight should correspond to the animal's weight determined from a scale.

The velocity fluctuations of the CM ($v$) are calculated by integration of each acceleration component determined from the force platform:

$$v_i(t) = \int_{t=0}^{t} a_i(t) + \mathrm{const}_i \tag{3}$$

where $i$ equals 1, 2, 3; $t$ is time; and $\mathrm{const}_i$ is the animal's average velocity in the $i$th direction.

The integration constants must be determined from the boundary conditions of the system. The average height of an animal's centre of mass varies little from stride to stride. Thus, the average vertical velocity ($\overline{v_3}$) after an integral number of strides is zero, a is the average lateral velocity ($\overline{v_1}$) if the animal moves in a straight line. The average horizontal speed ($\overline{v_2}$) has to be measured separately. This can be done by measuring the time that it takes an animal to cover a given distance. If the track is long enough reasonable estimates can be obtained by just measuring the time from the first step on and off the forceplate track of known length (error is about 10%; ref. 12). Speed can also be calculated from the time an animal needs to cover the distance between two photocells (13). However, errors can result in this estimate if different parts of the animal's body cross each of the two photocells. Also the distance between the photocells usually does not consist of an integral number of strides.

Changes in the kinetic energy of the CM ($E_{\mathrm{kin}}$) can be calculated from the instantaneous velocities of the animal's CM:

$$E_{\mathrm{kin},i}(t) = \frac{m}{2} v_i(t)^2. \tag{4}$$

Changes in potential energy ($E_{pot}$) of the CM can be calculated from the vertical displacement ($s_3$), which is obtained in turn by integration of the changes of the vertical velocity of the centre of mass:

$$s_3(t) = \int_{t=0}^{t} v_3(t) + \text{const.} \tag{5}$$

The corresponding integration constant must be zero, since the average vertical displacement over an integral number of strides is unchanged. Changes in the potential energy of the animal's CM equals:

$$E_{pot}(t) = mg\, s_3(t). \tag{6}$$

The total external energy of the CM ($E_{ext}$) is calculated from the sum of kinetic and potential energy components:

$$E_{ext}(t) = E_{pot}(t) + \frac{m}{2} \sum_{i=1}^{3} v_i(t)^2. \tag{7}$$

Minimum estimates of mechanical work have been calculated by summing the positive increments of total external energy ($+\Delta E_{ext}$) for each stride (7, 12, 14, 15). However, see Section 5.1 for the assumptions of this calculation, since other energy components may not be accounted for by the CM approach.

## 2.2 Measurement of external forces of the CM

### 2.2.1 Design features of force platform for mechanical work measurements

Force platforms are most often used to measure the ground reaction forces required to calculate the energetic fluctuations of the CM (see Chapter 3.2). Ground reaction forces must be measured in at least the vertical and horizontal direction. In many lower vertebrates and arthropods, the lateral force component is also essential.

If the mechanics of constant, average speed locomotion are desired, then some criteria must be set to eliminate trials in which the animal is generating net accelerations or decelerations. Typically, this is determined from the sums of positive and negative changes in velocity. Trials in which the difference of these sums is less than 15–25% have been accepted as a constant, average speed (12, 14). To judge whether animals are moving without net accelerations or decelerations, the force platform must be long enough to measure at least one complete stride with the whole animal on the platform. This requires a long force platform. To keep the resonant frequency of the force platform significantly above the highest frequencies of the investigated force signal, the platforms must be stiff and of low weight. On the other hand, the

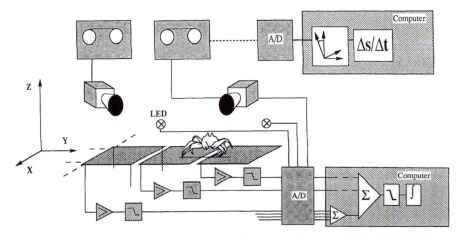

**Figure 1.** Experimental set-up. The signals of each force component in the corners of the platform are amplified and filtered. Subsequently these signals are digitized, summed for all corners of the platform and all platforms, filtered and integrated. Photocells can be used to provide an estimate of speed. Two high-speed cameras can collect information on the localization of body segments in three-dimensional space. After transformation of the respective coordinates, speed and angular velocities of the segments can be obtained by differentiation.

sensitivity of the plate is proportional to its compliance. Cavagna (11) and Heglund (13) solved this problem by building a track of small force platforms placed in series (*Figure 1*). The independent signals from the platforms are summed to calculate total force. The disadvantage of this design is that it becomes increasingly difficult to build and to maintain a number of platforms. By using semiconductor strain-gauges at the force-transducing support arms of the platform, stiffness of the platforms can be increased, and the number of platforms reduced (12).

### 2.2.2 Data processing

Data can be processed by analogue (11) or digital techniques (16). The speed and capacity of computers and their widespread use make the latter the more promising option. In digital processing, flexibility is guaranteed and the original data can be used for other analyses.

One problem that arises from the collection of force data is baseline drift. High speed taping for kinematic analysis requires bright illumination which in turn heats up the force platform components, causing drift in the voltage output of the electronic circuits. The drift in the baseline of the signal can be corrected by sampling a series of points before and after the time that the animal steps on the platform. These values can be used to provide a linear approximation of the baseline drift which can then be removed at any point in time during the recording of force (16).

Another major difficulty with the force signals is noise due to 'ringing' of the platform associated with its natural frequency. A variety of filters are available to reduce the 'ringing' noise in the force data (9, 17). Initial analogue filtering prior to digital processing may be necessary to avoid aliasing. Problems resulting from phase shifts using analogue filters may be considerable (see Chapter 3, Section 3.4), and it is advisable to use cut-off frequencies as close to the sampling frequency as possible. For further processing, digital filtering is often preferable because original data can be passed through a variety of filters to avoid distortion. Whatever the digital filter selected, it should be symmetrical to avoid phase shifts, it should be steep at the cut-off frequency, and the sum of the coefficients should be one, in order to maintain signal amplitude constant.

### 2.2.3 Signal integration

The calculation of potential and kinetic energy requires integration of the force data. Each integration process suppresses the high-frequency components and thus, results in effective filtering of the velocity and displacement data. As stated by Cavagna (11), analogue integration of the force data and the calculation of the velocities can be advantageous as 'the integration process abolishes the interference caused by the vibrations of the plate'. Simple analogue circuits for integration can be built easily (see ref. 13, and ref. 17, p. 121), but there are a number of pitfalls which diminish the apparent advantage of analogue integration. In particular, the signal must be extremely stable and any offset, including errors in the subtracted weight, must be very small as they are integrated twice to determine the changes in potential energy.

Once again, digital solutions to noise reduction and integration are preferable. The simplest approach to integration is the trapezoidal rule. Here, each area between successive data points is approximated by a trapezoid:

$$\int_{t=0}^{t} f(t) \, dt = f_{sample^{-1}} \frac{1}{2} [y(0) + 2y(1) + 2y(3)$$
$$+ \ldots + 2y(N - 1) + y(N)] \tag{8}$$

where $N$ is the sample size and $f_{sample}$ is the sampling frequency. This represents a fast approach which can even be implemented with macros linked in electronic spreadsheets (12). The error from this calculation can be considerable, but is reduced by using a high sample frequency and/or by applying more elaborate integration techniques (for example, Romberg's method in ref. 18, p. 287).

## 2.3 Calibration and check of hardware, electronic circuits, and software

To test the operation of all circuits and software used to calculate the energy changes of the centre of mass, a sample test should be performed that provides a direct comparison between experiment and theory. A convenient tool for this purpose is a pendulum of appropriate mass placed on the force plate (*Figure 2*; ref. 12). The forces exerted by the pendulum are generated by the swinging mass and transmitted by the string to the support of the pendulum and thus to the platform. If $\theta$ describes the angular deflection of the pendulum (length, $l$, and mass, $m$), then the force in the string ($F_{string}$) is the sum of the centrifugal force and the component of gravity in the direction of the string:

$$F_{string} = m \, \ddot{\theta}^2 \, l + m \, g \cos \theta. \tag{9}$$

For the horizontal and vertical component of the ground reaction force:

$$F_1 = \sin \theta \, F_{string}; \quad F_2 = \cos \theta \, F_{string}. \tag{10}$$

For small displacements $\theta$ is a harmonic function of $t$:

$$\theta = \theta_0 \sin \omega t \tag{11}$$

with $\omega^2 = \dfrac{g}{l}$.

For small angles, equation (9) can be integrated analytically to obtain velocities, displacements, and energies; for large angular deflections numerical integration is necessary. This results in theoretical values which can be compared with experimental data. Representative experimental calibration curves for acceleration, velocity, and energy fluctuations of a pendulum are shown in *Figure 2B*. (The pendulum should show near 100% energy recovery.)

# 3. Measurement of internal work

Body appendages can move relative to one another in such a way that the body's CM is not affected. Nevertheless, muscle force and thus metabolic energy are necessary to power these movements. Internal work relative to the CM (19) and total work (8) have been estimated by summing the energy changes from each segment (that is, in a linked segment model) over an integral number of strides. This technique of using kinematic data or a combination of kinematic and external force data is commonly referred to as *'inverse dynamics'*. Joint moments and joint reaction forces are estimated from external movements and forces. This is the reverse order of what happens in the body, since muscles actually produce the joint moments and forces which result in the movement of segments.

## 3.1 From kinematics to internal work

The body of an animal can be approximated by a series of linked segments. The location of the joint—or, more accurately, the instantaneous axis of rotation and the position of the CM must be determined for each body

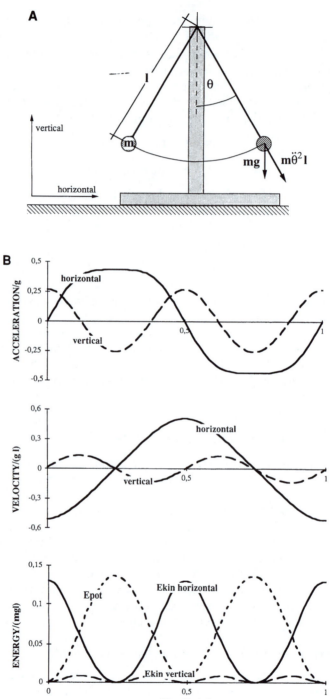

segment relative to the segment markers used in video taping (see below). The movement of each segment can then be described by a translation of the segment's CM and a rotation around the CM (*Figure 3*). The inertia of each segment is proportional to its mass for translational displacements and to its moment of inertia for rotational movements.

Changes in potential energy ($E_{pot,k}$) can be calculated from the vertical translation or displacement ($s_{3,k}$) of the centre of gravity of each body segment ($k$) where:

$$E_{pot,k}(t) = m_k \, g \, s_{3,k}(t). \tag{12}$$

By differentiation of the displacements of the CM of the individual segments, changes in their velocity ($v_{i,k}$) and thus the changes in translational kinetic energy ($E_{kin,t}$) can be calculated:

$$E_{kin,t,i,\,k}(t) = \frac{m_k}{2} \, v_{i,k}^2(t). \tag{13}$$

By differentiating the angle of rotation ($\theta$) with respect to time, the rotational velocity ($\omega$) of each segment can be calculated. Provided that the three-dimensional moments of inertia ($\mathbf{I}$) of the segment are available (see Section 3.2 below), the rotational energy of the segment ($E_{kin,r}$) can be determined:

$$E_{kin,r,i,k}(t) = \frac{1}{2} \sum_{j=1}^{3} \omega_{i,k}(t) \, I_{ij,k} \, \omega_{j,k}(t) \tag{14}$$

where $i \, (= 1, 2, 3)$ and $j$ denotes the axes of the Cartesian coordinate system. If the rotation takes place only around the $x$-axis ($i = j = 1$), then the rotational energy can be written:

$$E_{kin,r,1,k}(t) = \tfrac{1}{2} \, \omega_{1,k} \, I_{11,k} \, \omega_{1,k} = \tfrac{1}{2} \, I_{11,k} \, \omega_{1,k}^2. \tag{14a}$$

If the actual rotation has two components (for example, $x$ and $y$; $i, j = 1, 2$), then the rotational energy becomes:

$$E_{kin,r,1,k}(t) = \tfrac{1}{2} \, [\omega_{1,k} \, I_{11,k} \, \omega_{1,k} + \omega_{1,k} \, I_{12,k} \, \omega_{2k}] \tag{14b}$$

where $I_{12}$ denotes the element of the tensor of inertia that considers the influence of the movement around $y$ ($j = 2$) on the first energy component ($i = 1$). If the movement is planar and the segments have a parallel plane of symmetry, equation (14) simplifies to:

$$E_{kin,r,i,k}(t) = \tfrac{1}{2} \, I_{i,k} \, \omega_{i,\,k}^2(t). \tag{15}$$

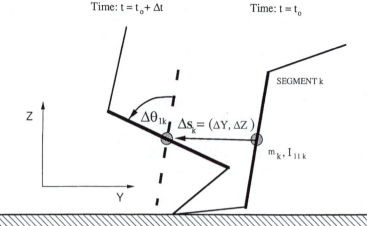

**Figure 3.** Notation for the calculation of internal energy. The leg of the animal is moving from right to left in the *yz*-plane. Between two frames (time interval = $\Delta t$) the CM of the *k*th segment ($m_k$ = mass; $I_{11}$ = moment of inertia for rotation around the CM in the *yz*-plane) is translated by $\Delta s_k$ ($\Delta y_k$, $\Delta z_k$) towards the position marked by the dashed line and rotated by $\Delta\theta_{lk}$ from the dashed line to its final position. From the change in position, the velocity of translation ($v_k = \Delta s_k \, \Delta t^{-1}$) and the angular velocity ($\omega_k = \Delta\theta_{lk} \, \Delta t^{-1}$) can be determined. Velocity changes allow calculation of kinetic energy. Indices of angular changes denote the axis of rotation which is perpendicular to the instantaneous plane of movement (for example, parallel to $x = x_1$).

The mechanical energy contributed by each moving segment ($E_k$), then, is the sum of the respective kinetic (translational and rotational) and potential energies at each instant in time:

$$E_k(t) = E_{\mathrm{kin},k}(t) + E_{\mathrm{pot},k}(t). \tag{16}$$

Considering one plane ($i,j$) and planar symmetry of the body this yields:

$$E_k = \tfrac{1}{2} \, m_k \, (v_{i,k}^2 + v_{j,k}^2) + \tfrac{1}{2} \, I_{ii,k} \, \omega_{i,k}^2 + m_k \, g \, s_3. \tag{17}$$

The total internal energy ($E_{\mathrm{int}}$) of the moving animal can be calculated by adding the contributions from every segment at each instant in time:

$$E_{\mathrm{int}}(t) = \sum_{k=1}^{n} E_k(t). \tag{18}$$

Once again, there are numerous ways to calculate internal energy from the same data set depending on the assumptions made about the linked system (see Sections 5.1 and 5.2).

## 3.2 Determination of the CM and moment of inertia of body segments

Whereas the instantaneous axis of rotation can be determined from kinematic data, the CM and the moment of inertia must be determined for each segment independently.

The simplest approach used to determine the CM is to freeze and cut the body into segments. The CM is then determined by photographing or videotaping each segment as it is suspended from various axes. The CM is marked by the common crossing of all vertical lines drawn from the respective points of suspension. The CM of a multi-segmented system is (9):

$$s_i = \frac{1}{M} \sum_{k=1}^{n} m_k \, s_{i,k} \tag{19}$$

where $s$ is the position of the CM of a multi-segmented body, $s_{i,k}$ is the position of the CM of a segment $(k)$, $m_k$ is the mass of a segment $(k)$, $n$ is total number of segments and $M$ is total mass $= \Sigma_{k=1}^{n} m_k$.

A similar segmental approach can be used to determine the moment of inertia. However, it is important to note that the moment of inertia of a segment rotating in three-dimensional space can be complicated to determine. Being three-dimensional, the moment of inertia is a tensor. It is characterized by three quantities $(I_{11}, I_{22}, \text{and } I_{33})$, if the principal axes of the tensor are known. Otherwise, six quantities are needed for a complete description:

$$\mathbf{I} = \begin{pmatrix} I_{11} & -I_{12} & -I_{13} \\ -I_{21} & I_{22} & -I_{23} \\ -I_{31} & -I_{32} & I_{33} \end{pmatrix} \tag{20}$$

where $I_{ij} = I_{ji}$.

If the body under investigation has no symmetry and the main axis of the moment of inertia are not known, then the six independent experimental estimates are necessary to quantify this tensor using a linear system of equations and transformation of coordinates. After transformation of the coordinate system to the principal axes of the segment, the tensor contains only the three diagonal elements:

$$\mathbf{I} = \begin{pmatrix} I_{11} & 0 & 0 \\ 0 & I_{22} & 0 \\ 0 & 0 & I_{33} \end{pmatrix}. \tag{21}$$

Frequently, the segment can be approximated reasonably by a body with two perpendicular planes of symmetry with the principal axes of the tensor perpendicular to these planes (*Figure 4*).

Modern tomographic methods allow a precise three-dimensional estimate of the geometry of body segments (Brüggemann, pers. comm.). Even the position of more dense skeletal structures can be determined. These techniques are expensive, but have the advantage of not requiring the sacrifice of an animal. If the density and the shape of segments of the specimens are

**85**

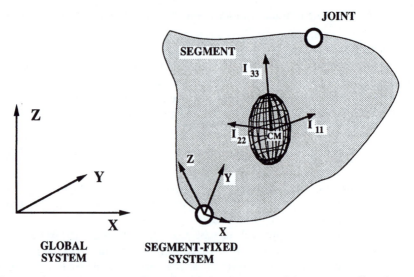

**Figure 4.** In a three-dimensional representation the tensor of the moment of inertia can be visualized by an ellipsoid. The lengths of the principle axes of this ellipsoid to $I_{11}$, $I_{22}$, $I_{33}$, respectively, and are sufficient to characterize inertia of any three dimensional body. The axes of the segment-fixed coordinate system (X,Y,Z) are frequently chosen to lie parallel to the axes of rotation defined by the construction of the joint. If the orientation of the segment-fixed system is different from the orientation of the principle axes and the segment does not rotate around one of the main axes, then the tensor must be rotated into the segment-fixed system and six quantities are necessary to describe inertia.

known, then the moment of inertia can be calculated by computing the corresponding volume $(V)$ integrals of each of the tissues in the body segment:

$$I_{ii} = \rho_V \int (x_j^2 + x_l^2) \, dV, \; I_{ij} = \rho_V \int x_i x_j \, dV \tag{22}$$

where $x_{i,j,l}$ are the Cartesian coordinates of points within the volume of the tissue, the origin being located at the CM of the segment. Density is represented by $\rho$; $m = \rho V$; $i,j,l = 1,2,3$; and $i \neq j \neq l$  for example:

$$I_{11} = \rho_V \int (x_2^2 + x_3^2) \, dV \text{ and } I_{12} = \rho_V \int x_1 \, x_2 \, dV. \tag{22a}$$

In many cases complicated shapes of segments can be approximated by a stack of geometrical elements $(m)$ such as cylinders, truncated cones, etc., with known moments of inertia. The moment of inertia of the whole segment relative to rotations around the CM can be calculated by:

$$I_{ii} = \sum_m [I_{ii,m} + m_m (x_{j,m}^2 + x_{l,m}^2)] \tag{23}$$

$$I_{ij} = \sum_m [I_{ij,m} + m_m (x_{i,m} x_{j,m})] \tag{24}$$

where $i,j,l = 1,2,3$ and $i \neq j \neq l$ (see equation (22)) $x_{i,m}$ is the difference in the coordinates of the CM of the total segment and the CM of elements of the stack (20).

Experimentally, the moment of inertia around any axis of rotation can be determined by letting the object oscillate as a physical pendulum:

$$I = \tau^2 \frac{m\,c\,g}{4\,\pi^2} \tag{25}$$

where $\tau$ is the period of the pendulum, $m$ is the mass of the segment, and $c$ is the distance from the axis of rotation to the CM.

Often, however, the segment of interest rotates about an axis other than its CM. The moment of inertia around this axis ($I$) can be calculated by applying the parallel axis theorem:

$$I = I_0 + mc^2 \tag{26}$$

where $I_0$ is the moment of inertia around the CM (as determined above).

## 3.3 Data processing

Cinematographic techniques and kinematic analyses are described in detail elsewhere in this volume (see Chapter 3). In general, because the determination of segmental energies requires the calculation of velocities from displacement data, the application of fast and highly accurate cinematographic techniques are a necessity. The major problems encountered include digitizing error, unwanted movement of skin markers with respect to underlying skeletal structures and inadequate sampling rate.

# 4. Calculation of joint and muscle work and power

Muscles can both generate and absorb energy. Muscles that generate energy do positive work, whereas 'negative' work is 'done on' muscles that absorb energy. If both of these functions are important to the musculoskeletal system that is to be analysed, then in principle energy changes can be calculated from the net moments or torques ($T$) developed at each joint ($k$; see also Chapter 3; refs. 3 and 21). The rate of work done by or to muscles varies with time. Instantaneous muscle power at a joint ($P_m$; ref. 22) is the product of the net moment generated by the muscles at each joint ($T_k$) and joint angular velocity ($\omega_k$):

$$P_m = T_k\,\omega_k \tag{27}$$

The calculation of net joint moments requires combined ground reaction force data obtained from a force platform and kinematic data from high-speed taping (Chapter 3). Ground reaction forces, segment masses, and moments of inertia are used in a free body diagram analysis to calculate the net moment at the joint (depending on the importance of the inertial component; Chapter 3, Section 4).

Joint angular velocity at any instant is obtained directly from the kinematic analysis (Chapter 3, Section 3.5).

Total energy ($E_{tot}$) or work done is determined by integrating $P_m$ over an integral number of strides for each joint:

$$E_{tot} = \sum_k \int |T_k \, \omega_k| \, dt. \tag{28}$$

In these calculations it is essential to be consistent in defining the direction of the joint moment (for example, counter-clockwise moments are positive, whereas clockwise ones are negative). To estimate the total positive and negative work done, it is important to sum both negative (energy absorbing) and positive (energy producing) areas of the $P_m$ versus time function separately (this is the reason for the absolute value sign in equation (28), but see Section 5.2 below).

# 5. Evaluation of techniques used to examine the mechanical work in terrestrial locomotion

## 5.1 Energy exchange within segments and transfer between segments

In his initial attempts to estimate the work done during terrestrial locomotion, Fenn (4) calculated the increases in potential and kinetic energy of a body's segments. Summation of the increases in segment energy, however, can lead to an overestimate of the work done, if energy exchanges within and between segments are significant. Consider a limb that functions as an ideal pendulum. Once energy is put into the system, no additional energy input would be required to swing the limb, because all the potential energy would be exchanged with kinetic energy. Artificially high values of mechanical work (termed 'pseudo-work') would result, if the potential energy increases were simply added to kinetic energy increases.

Cavagna and others (5, 6, 7) demonstrated the importance of energy exchange and recovery during walking by calculating the energy fluctuations of the CM from ground reaction forces (that is, external work or CM approach). During walking in bipeds, quadrupeds, and even eight-legged crabs, potential and kinetic energies of the CM are out of phase by about 180 degrees much like an egg rolling end over end or an inverted pendulum (*Figure 5*; refs. 7, 14, 16). Energy lost in the kinetic form as the body slows can be recovered and used to raise the body and increase potential energy. Therefore, if the potential and kinetic energy of an animal using a pendulum-like energy exchange mechanism are summed at each instant, then the fluctuations of total, external energy ($\Delta E_{ext}$) will be reduced. The possible degree of energy exchange or recovery can be quantified as (7):

$$\text{recovery } [\%] = 100 \, \frac{|\Delta E_{pot}| + |\Delta E_{kin}| - |\Delta E_{ext}|}{|\Delta E_{pot}| + |\Delta E_{kin}|} \tag{29}$$

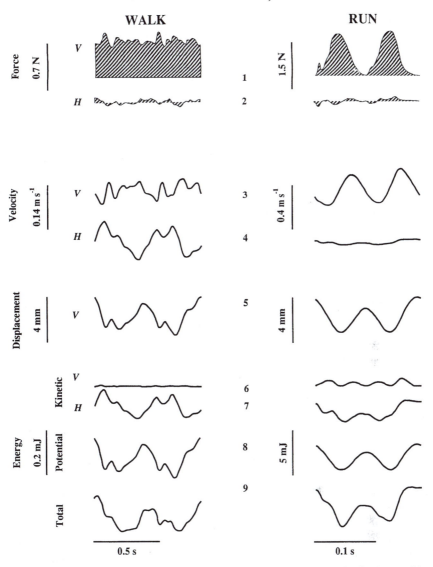

**Figure 5.** Example: Fluctuations of external energy in the ghost crab. During walking potential and kinetic energies are out of phase allowing for energy exchange. During running both energies are in phase energy exchange is small but energy storage is facilitated. (Adapted from ref. 16.)

where $\Delta E_{pot}$ and $\Delta E_{kin}$ are the changes in potential and kinetic energy, respectively. Recovery would be 100% for an ideal pendulum. Surprisingly, recovery in large mammals during walking can reach 70% (7).

In contrast to walking, potential and kinetic energy fluctuations of a body's CM have been shown to be in phase during running in bipeds, quadrupedal mammals, crabs, and insects (7, 14, 23). Potential and kinetic energy fluctuations of the centre of mass add to result in large changes in the total, external energy (Fig. 5; ref. 7). The system operates similar to a spring–mass system and pendulum-like energy recovery decreases. However, potential and kinetic energy can be stored temporarily as elastic strain energy in tendons upon landing and then returned to the centre of mass upon take-off. Elastic strain energy storage has proven to play an important role in large mammals, such as the kangaroo, where as much as 50% of the mechanical energy can be stored and returned (24).

Analysing the energy fluctuations of the CM has lead to the development of spring–mass models of running (25, 26) as well as inverted pendulum models of walking (5, 7, 27). Although the CM approach has been invaluable in designing simple models of terrestrial locomotion, there are several conditions in which mechanical work estimates obtained by this method could differ significantly from actual work output:

- the body's CM may not reflect all the changes in energy of each segment (8)

- simultaneous increases and decreases in the energy of segments moving reciprocally and relative to the movement of the CM are not measured

- energy changes in the flight or aerial phase are not included in the estimation

- the CM approach does not account for energy generation and absorption at each joint (9)

Energy exchange is possible, not only for the CM, but for each body segment (8). Moreover, energy can be transferred between adjacent segments (ref. 4; 'whip effect'). Winter's (8) estimation of total body work (that is, internal work) includes the possibility of energy exchange within segments and between adjacent segments. Estimates of internal work will vary depending upon the assumptions made concerning energy exchange.

In general, the lowest estimates of work are obtained when complete energy exchange is hypothesized. If complete exchange is assumed within a segment, then the energy of a segment can be calculated by simply summing the energies (that is, potential, kinetic, and rotational) at each instant in time. If complete exchange is assumed between adjacent segments, then the energy of the body can be calculated by summing the instantaneous segment energies as shown above. Internal work will be the greatest when no energy exchange is hypothesized (4, 28). To calculate internal work with no exchange, the absolute value of the energy changes over time is summed.

Intermediate values of internal work result from assumptions of partial energy exchange. In a comparative study of terrestrial locomotion, Fedak *et al.* (19) calculated total internal energy (taken to be primarily kinetic energy) by assuming exchange within and between limb segments, but not between the limbs and body or other limbs. In other words, these workers assumed no exchange between internal work done by the limbs relative to the CM and external work done on the CM. However, energy exchange between internal and external energy can occur during the stance phase. Consider a stick rotating around an axis through it's centre of gravity. If such a stick touches the ground, then part of its rotational energy is transferred into kinetic energy of the CM, which can be easily measured from the generated ground reaction forces. This type of interaction is used (in reverse) by gymnasts to somersault and simple summation of the positive increments of the fluctuations of internal and external energy would lead to an over-estimation of internal work. Heglund *et al.* (29) determined the maximum error due to neglected exchange during steady locomotion to be within about 30% of total mechanical energy (such as for a chipmunk and a dog). For large animals, such as a horse, values of internal work derived from partial versus complete energy exchange could differ by more than a factor of two. Williams and Cavanagh (30) found that total work done during human locomotion can vary by 75% depending on the assumptions of energy exchange.

Regardless of the assumptions made concerning energy exchange, total work can be estimated by either summing the positive increases in total energy over an integral number of strides or by summing the absolute value of the changes. The former ignores the fact that absorbing mechanical energy costs metabolic energy and gives a minimum estimate of mechanical work. The latter estimation gives equal weight to both positive and negative work. A third possibility, and perhaps the most reasonable approach used by investigators, is the weighting of negative work to be $\frac{1}{3}$ to $\frac{1}{5}$ the energy cost of positive work (3, 30). The decision of if or how to include negative work can alter the total work output significantly, since an equal amount of positive and negative work is done during constant, average-speed locomotion. Consequently, it is important to make clear the assumptions adopted to compute total energy changes of the animal's body.

Recognition of energy exchange within and between segments has led to a better understanding of energy conservation and an improved estimate of total work. The total work estimate for human walking can be 16–40% greater than that calculated from the CM approach, if the energy changes of each segment are included (8). Internal work can be of the same magnitude as the external energy calculated from energy changes in the CM at the highest sustainable speeds used by quadrupedal mammals (29). However, it is important to note that the simple summation of segment energies at each instant can also underestimate the generation and absorption of energy at different joints (9). Because significant amounts of energy can be generated and absorbed simultaneously at different joints, estimations of work in terrestrial locomotion may be improved by examining individual joints.

## 5.2 Energy production and absorption

The generation and absorption of energy at a joint can be calculated from the net joint moment and joint angular velocity (see Section 4). Instantaneous

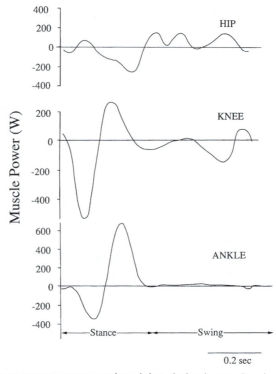

**Figure 6.** Summary power patterns at three joints during human jogging. (Adapted from ref. 10). Each plot shows the stance phase followed by the swing phase. The end of the stance phase is marked by the ankle's decrease of power to zero from the large maximum of power generation.

muscle power curves from a study on human jogging (*Figure 6*; ref. 10) show the advantages of using this technique. Muscle power is positive for concentric (shortening) muscle contractions in which the net moment and the joint angular velocity are of the same polarity. Energy is absorbed in eccentric (lengthening) contractions when muscle power is negative. The hip has relatively low power levels compared to the knee and ankle, probably functioning to stabilize the trunk. Surprisingly, the knee extensors do not generate energy to swing the leg (that is, no positive power). Instead, the knee absorbs energy from the swinging leg. Knee extensors absorb nearly four times as much energy as they produce. The ankle extensors clearly generate the majority of energy, producing three times as much positive work as the knee extensors.

If data on energy generation and absorption at joints are combined with energy transfer analysis, an even more complete picture of energy flow can be drawn. Energy transfer, generation, and absorption at a joint can be calculated from joint reaction forces and moments on leg segments (22, 31, 32). The power between segments acting across joints ($P_b$) is:

$$P_b = F_k \cdot V_k \tag{30}$$

where $F_k$ is the reaction force vector at joint b and $V_k$ is the velocity vector of the centre of the joint. Total power for a segment is the sum of the joint power ($P_b$) and the muscle power ($P_m$; see equation (27)); the only difference being that the muscle power is calculated using the absolute angular velocity of the segment as opposed to the joint angular velocity. *Figure 7* (from ref. 32) shows data from a power generation, absorption and exchange analysis conducted on human walking. Joint power is shown by an arrow crossing through the joint centre, whereas muscle power is represented by arrows around each joint on the side where energy is flowing. During push-off at the end of the support phase there is 533 W of power produced by the Achilles tendon. 65 W of energy is transferred to the foot (energy increase), but the majority of the energy (469 W) flows upward from the foot through the ankle joint. 108 W of energy continues upward throughout the knee joint, with little participation by knee muscles. Only 23 W of energy flows into the trunk. During weight acceptance at the beginning of ground support, 244 W flow out from the trunk and across the knee. Nearly 164 W is absorbed by the knee extensors and 33 W (that is, 145–108 W) is absorbed by the ankle dorsiflexors.

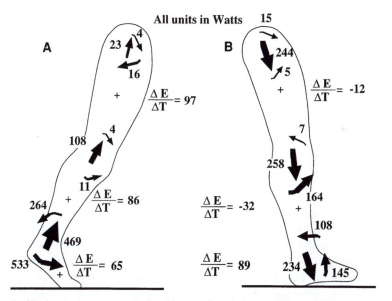

**Figure 7.** Mechanical power analysis of human leg during walking at (**A**) push-off and (**B**) weight acceptance. The joint power is shown by an arrow crossing through the joint centre. The muscle power is represented by arrows around each on the side where energy is flowing. Power is in watts. (adapted from ref. 32.)

**Table 1.** Powerflow for a single joint. For a positive (counterclockwise) angular velocity of the proximal segment 1 ($\omega_1 > 0$), the power can be transferred from segment 1 to the distal segment 2 ($\rightarrow$) via the extensor of the joint. Complete transfer (tr.) occurs if the angular velocities are similar. If $\omega_1 > \omega_2$, energy is absorbed in the extensor (abs.), the muscle contracts, i.e. power is generated (gen.). For $\omega_1 < 0$, the flow is reversed and gen. and abs. must be exchanged in this table.

|  | Direction | $\omega_1 > \omega_2$ | $\omega_1 = \omega_2$ | $\omega_1 < \omega_2$ |
|---|---|---|---|---|
| Extensor | $1 \rightarrow 2$ | tr. and abs. | complete tr. | tr. and gen. |
| Flexor | $1 \leftarrow 2$ | tr. and gen. | complete tr. | tr. and abs. |

Given the absolute angular velocity of a segment and the joint moment, it is possible to describe the work relationships between two segments connected by an active muscle (21, 31). *Table 1* shows the function of muscle and the direction of power flow for a single joint muscle performing either flexion or extension. If one segment is fixed while the other is in motion, then the muscle can be either generating or absorbing energy. If $\omega_1 > \omega_2$, then the muscle is lengthening and absorbing energy while transfer is taking place. If $\omega_1 < \omega_2$, then the muscle is shortening and generating energy while transfer occurs. If $\omega_1 = \omega_2$ (that is, isometric contraction), then energy is only transferred by muscles from segment 1 to segment 2.

The determination of energy generation, absorption and transfer at the joints provides considerably more information about muscle work than most other techniques; however, even this technique cannot deal with the problems of co-contractions of antagonist muscles, isometric contractions, and elastic strain energy storage.

## 6. Muscle work

Perhaps, the most difficult estimate that one can attempt is a true estimate of a muscle's efficiency (that is, mechanical energy output/metabolic energy input) during terrestrial locomotion. A discussion of muscle efficiency is outside the scope of this chapter, but we will point out the several well-known factors that may preclude linking the above mechanical work estimates for terrestrial locomotion directly to muscle function.

Muscles that operate nearly isometrcally generate and absorb little mechanical energy. This would include muscles generating the force necessary to hold up the body or limbs against gravity at rest. Muscles that function isometrically can demand significant metabolic fuel consumption. In fact, considerable evidence exists to suggest that the metabolic cost of pedestrian locomotion is largely determined by the cost of this force production (2, 33).

Not only can individual muscles operate isometrically, but antagonistic muscles actively contracting against one another generate no mechanical work and yet consume considerable metabolic energy to maintain isometric tension. In general, the effect of co-contraction is considered to be small during locomotion, but this requires further analysis. According to Alexander and Vernon (24) 15% of total concentric muscle work for a kangaroo hopping at 5.5 m/sec include work done against antagonists. The degree of co-contraction during walking in humans has been estimated to be 24% (34).

In our discussion of mechanical work, we referred to muscle force or power. To be more accurate, we were describing musculotendon force or power. Elastic strain energy storage in tendons and apodemes may significantly alter the actual work a muscle must do if the stored energy can be returned to the segments. Estimates of the energy saved by muscles due to elastic strain energy storage vary considerably. In humans mechanical power output can vary by 40% depending on the amount of elastic strain energy storage (30). Kangaroos can save as much as 59% (24). Elastic strain energy storage in small mammals may be more limited because they have relatively thicker tendons (35). The calculation of tendon strain energy is found in Chapter 3, Section 4.2.

Even if one were able to estimate isometric force production, muscle co-contraction and elastic strain energy recovery at a joint, a major difficulty may still remain. Because more than one muscle may have a similar function at a joint, individual muscle work estimates are often indeterminant (see Chapter 3, Section 4). *In vivo* recordings of muscle force, EMG quantification, and computational methods that minimize or assume equal stress or power output have all been used to address this difficulty. Further study directed toward integrating muscle function with limb performance is obviously required and will be a promising area of research in the future.

Finally, although calculation of work in terrestrial locomotion is challenging, it has led to a better understanding of musculoskeletal function generally. By continuous revision and examination of assumptions, considerable progress has been made in linking muscle function with mechanical work. For the future, we argue that one of the strongest approaches to uncover the general biomechanical rules of terrestrial locomotion is the comparative method which takes advantage of natural experiments, involving animals that possess a diversity of musculoskeletal systems and live in many different environments.

## Acknowledgement

Thanks are due to A. Biewener for his numerous suggestions.

## References

1. Schmidt-Nielsen, K. (1972). *Science,* **177,** 222.
2. Full, R. J. (1991). The concepts of efficiency and economy in land locomotion. In: *Concepts of efficiency, economy and related concepts in comparative animal physiology* (ed. R. W. Blake) Cambridge University Press, New York. (In press.)

3. Margaria, R. (1976). *Biomechanics and energies of muscular exercise*. Oxford University Press, Oxford.
4. Fenn, W. O. (1930). *Am. J. Physiol.,* **92**, 583.
5. Cavagna, G. A., Thys, H., and Zamboni, A. (1976). *J. Physiol.,* **262**, 639.
6. Cavagna, G. A., Kaneko, M. (1977). *J. Physiol.,* **268**, 467.
7. Cavagna, G. A., Heglund, N., and Taylor, C. R. (1977). *Am. J. Physiol.,* **233**, R243.
8. Winter, D. A. (1979). *J. appl. Physiol.,* **46**, 79.
9. Winter, D. A. 1990. *Biomechanics and motor control of human movement*. John Wiley, New York.
10. Winter, D. A. (1983). *J. Biomech.* **16**, 91.
11. Cavagna, G. A. (1975). *J. appl. Physiol.,* **39**, 174.
12. Full, R. J. and Tu, M. S. (1990). *J. exp. Biol.,* **148**, 129.
13. Heglund, N. C. (1981). *J. exp. Biol.,* **93**, 333.
14. Heglund, N. C., Cavagna, G. A., and Taylor, C. R. (1982). *J. exp. Biol.,* **79**, 41.
15. Full, R. J. and Tu, M. S. (1990). *J. exp. Biol.,* **148**, 129.
16. Blickhan, R., and Full, R. J. (1987). *J. exp. Biol.,* **130**, 155.
17. Horowitz, O. and Hill, W. (1980). *The art of electronics*. Cambridge University Press, Cambridge.
18. Fröberg, C-E. (1985). *Numerical mathematics. Theory and computer applications*. Benjamin/Cummings Co., Menlo Park, California.
19. Fedak, M. A., Heglund, N. C., and Taylor, C. R. (1982). *J. exp. Biol.,* **79**, 23.
20. Hatze, H. (1979). *TWISK 79; a model for the computational determination of parameter values of anthropomorphic segments*. Tegniese Verslag, Pretoria.
21. Aleshinsky, S. U. (1986). *J. Biomech.,* **19**, 287.
22. Quanbury, A. O., Winter, D. A., and Reimer, G. D. (1975). *J. Human Movement Studies,* **1**, 59.
23. Full, R. J. 1989. Mechanics and energetics of terrestrial locomotion: From bipeds to polypeds. In: *Energy transformation in cells and animals* (ed. W. Wieser and E. Gnaiger), p. 175. Thieme, Stuttgart.
24. Alexander, R. McN., and Vernon, A. (1975). *J. Zoo. Lond.,* **177**, 265.
25. Blickhan, R. (1989). *J. Biomech.,* **22**, 1217.
26. McMahon, T. A. and Cheng, G. C. (1990). *J. Biomech.,* **23**, 65.
27. Mochon, S. and McMahon, T. A. (1980). *J. Biomech.,* **13**, 49.
28. Norman, R. W., Sharratt, M., Pezzack, J., and Noble, E. (1976). *Biomechanics*. Vol. V-B, p. 87.
29. Heglund, N. C., Fedak, M. A., Taylor, C. R., and Cavagna, G. A. (1982). *J. exp. Biol.,* **97**, 57.
30. Williams, K. R. and Cavanagh, P. R. (1983). *J. Biomech.,* **16**, 115.
31. Robertson, D. G. E. and Winter, D. A. (1980). *J. Biomech.,* **13**, 845.
32. Winter, D. A. and Robertson, D. G. (1978). *Biol. Cybern.,* **29**, 137.
33. Kram, R. and Taylor, C. R. (1990). *Nature,* **346**, 265.
34. Falconer, K. and Winter, D. A. (1985). *Electromyogr. & Clin. Neurophysiol.,* **25**, 135.
35. Biewener, A. A. and Blickhan, R. (1988). *J. exp. Biol.,* **140**, 243.

# 5

# Aerodynamics of flight

ROBERT DUDLEY

## 1. Introduction

Flying animals comprise the most morphologically diverse and numerically abundant form of life in the terrestrial biosphere. Animal flight can be described at a variety of phenomenological levels, ranging from the biophysics and ultrastructure of flight muscle to the behaviour of migrating locust swarms. An aerodynamic analysis of flight entails description of wing and body motions, evaluation of the forces associated with these motions, and calculations of the concomitant power requirements during locomotion. Additionally, the implications of wing and body morphology for more general measures of flight capacity, such as characteristic airspeed and manœuvrability, can be assessed quantitatively.

Such a comparative biomechanical approach allows functional consequences of organismal design as well as potential physical constraints on flight performance to be identified. For example, only an understanding of the physical mechanisms of wing flapping will permit elucidation of the effects of wing size and shape on airflow patterns during flight. The mechanical power required to fly will in part establish the total energetic costs of locomotion, a quantity of immediate ecological significance. As technical exercises in elucidating structural function, aerodynamic analyses of animal flight can be of considerable interest even when devoid of reference to related aspects of biology. However, far greater insight into organismal design will result if biomechanical findings can be placed in ecological and evolutionary contexts.

Principal extant taxa of flying animals include the pterygote Insecta, bats, and birds. Theory and methods for studying vertebrate flight have recently been the subject of two book-length treatments (1, 2), and no attempt will be made here to repeat this detailed coverage. Instead, this chapter will emphasize the biomechanical analysis of insect flight. In most respects, the aerodynamic theory required for studying invertebrate and vertebrate flight are equivalent, but frequently both terminology and means of mechanical analysis differ substantially. This chapter will strictly follow the practices of contemporary research in insect flight aerodynamics. A symbols appendix is included to facilitate identification of important variables and parameters.

The organization of the chapter parallels the methodological approach to be followed by a prospective experimenter interested in studying the biomechanics of insect flight. Means of eliciting and controlling flight in insects are first discussed. The kinematics of flight (that is, motion of the wings and body) can then be described once flight or a reasonable approximation thereof has been attained. Using isolated wings and bodies, a variety of morphological, structural, and aerodynamic characteristics can be evaluated under static conditions. Given these and the kinematic data, aerodynamic forces and the associated mechanical power requirements of flight can then be estimated. The chapter concludes by discussing simple yet informative studies of comparative flight performance that would enhance our understanding of morphological diversification in the pterygote Insecta.

# 2. Getting insects to fly

The initial task in any study of insect flight is to elicit either wing flapping or actual free flight from the experimental subject. Three principal methods are used: tethered flight, free flight in chambers and enclosures, and natural free flight in the field.

## 2.1 Tethered flight

While some of the earliest studies of insect flight investigated free-flying insects, many contemporary studies have involved tethering. Typically in tethered flight, a rigid support is attached with adhesive to either the thorax or abdomen of the insect. Flexible tethers can also be used, in which case a loop of thread or monofilament is placed around the thorax or abdomen and tightened. Tethering is convenient because the insect can be positioned according to the experimenter's will, and wing flapping can usually be elicited simply by evoking the tarsal reflex (removal of support from the legs). The tethered insect can be placed in the jet of a wind tunnel, with the effective forward velocity either selected by the experimenter or by the insect itself. In this latter arrangement, thrust generated by the insect is measured electronically and used to control the tunnel airspeed via a feedback loop connected to the tunnel motor.

The resemblance of tethered flight to free flight in insects is open to question. If the tether is attached to the thorax, dynamics of thoracic deformation during wing flapping may be adversely affected. Extreme behavioural responses may inadvertently be elicited. The similarity of wing motions to those during free flight is unknown, and in several cases kinematics have been shown to differ significantly. For example, wing motions described for flies in tethered flight (3) differ substantially from those of free-flying dipterans (4, 5). Concomitantly, the aerodynamic forces and pitching moments produced by the insect were probably not equivalent to those during natural flight. Without knowledge of these forces and rotational moments, the potential similarity of tethered wing motions to those in free flight must be treated cautiously. If tethered preparations (with either rigid or flexible tethers) are to be used in biomechanical studies of insect flight, the minimal experimental criteria to be met are that wingbeat frequency and stroke amplitude (Section 3.2.1) closely approximate those during free flight. Ideally, vertical force production should be measured through the course of a wingbeat (Section 3.4), and the time-averaged value thereof shown to equal the body weight.

## 2.2 Free flight in the laboratory

In contrast to the uncertainties associated with tethered flight, experimental realism is not an issue for insects flying freely in chambers, large enclosures, or insectaries. Also, microclimatic factors such as air temperature, ambient air motion, and relative humidity can be readily controlled under such conditions. While hovering flight is most easily studied in small flight chambers or enclosures, analysis of forward flight is not so straightforward, as many insects flying freely will do so irregularly or discontinuously. Such behaviour is required if the goal is to analyse complex three-dimensional flight paths, but in general precludes the analysis of steady flight at constant velocities. Behavioural manipulations (for example, use of lighting cues) coupled with careful film analysis of trajectories, however, may identify limited periods of unaccelerated horizontal motion.

Alternatively, controlled free flight can be elicited from an insect that is placed in the working section of a wind tunnel [(6); and see *Figure 1a*]. Such an arrangement utilizes the optomotor response (orientation of insects towards moving optical cues) and appropriate lighting to direct and orient the flying insect (*Figure 1b*). Although difficult to implement experimentally, controlled free flight affords the highest quality and greatest opportunity for biomechanical analysis. Insect airspeed can be controlled externally, while camera configuration during filming is not constrained by behaviour of the insect. The feasibility of behaviourally manipulating insects to fly at constant speeds in wind tunnels has not been systematically investigated, but this technique will probably work best for those neopterous insects capable of high precision and manœuvrability during flight (for example, various Diptera and Hymenoptera).

### 2.3 Flight in nature

Studying flying insects in the field is the most biologically relevant approach for biomechanical studies, but also presents the greatest logistical difficulties. In most natural situations, it is impractical to directly follow a flying insect. Usually, flight can only be filmed and analysed when the insect flies predictably through given locations. Airspeed of the insect is difficult to assess because ambient air motions can only be determined approximately (see

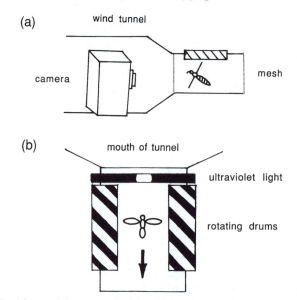

**Figure 1.** Method for studying controlled free flight of insects flying in the working section of a wind tunnel (modified from ref. 6). (a) Lateral view of tunnel depicting the flying insect and camera used for filming. (b) Vertical view of the ultraviolet light and rotating striped drums used to orient the insect during flight. The arrow indicates the direction of air flow.

Section 3.1). Subsequent capture of the insect following filming is often impossible in the field, and the morphological data necessary for biomechanical analyses must be obtained from pooled mean values for the species in question. Consequently, the often wide range of intraspecific morphological variation (and implications thereof for flight biomechanics) must be disregarded. These limitations aside, existing data on natural flight performance of insects are remarkably limited, and even rudimentary flight data taken in the field would contribute considerably to existing knowledge. Combining such data with detailed laboratory investigations of flight biomechanics will permit a broad overview of flight performance to be derived.

## 3. Flight kinematics

Filming of either tethered or free-flying insects is necessary to determine the kinematics of flight. The speed of flight is fundamental to any biomechanical analysis of flight. Wing and body motions relative to a coordinate system fixed in the insect can in many cases be derived from a single camera, while three-dimensional flight paths must be derived from two synchronized camera views. Patterns of airflow around the wings and bodies can also be visualized and in some cases quantified using tracer particles (Section 3.3).

## 3.1 Airspeeds

Knowledge of the airspeed of a flying insect is the most basic of kinematic information, but unfortunately can also be the most difficult to obtain accurately. In tethered flight, the speed of airflow past the insect can obviously be determined with ease, while flight speeds of free-flying insects 'tethered' remotely by optomotor cues in wind tunnels can similarly be manipulated and measured at will. Flight speeds in enclosures or screened insectaries (where ambient air motion is negligible) can usually be determined photographically by measuring the number of body lengths travelled per unit time, provided that the flight path is nominally orthogonal to the optical axis of the camera. As mentioned previously, filming of insects in the field usually occurs when the magnitude and direction of the prevailing winds are unknown, and use of the insect's groundspeed alone for aerodynamic analyses is inappropriate. However, there may be circumstances for which windspeeds are negligible (for example, filming early in the morning or in sheltered locales), in which case the groundspeed closely approximates the true airspeed.

A new experimental method permits direct airspeed measurements on insects flying over a body of water (7, 8). The investigator follows an insect in a small motorboat such that trajectories of the insect and boat are nominally parallel. A unidirectional anemometer (for example, TSI #1650, St Paul, MN) is held laterally from the prow of the boat such that the probe is in the immediate vicinity and at the same height as the flying insect (*Figure 2*). Airspeed is thus measured directly, and no assumptions concerning ambient wind are necessary. Repeated measurements of airspeed are possible if the insect can be followed for an extended period of time. Conveniently, the same insect can then be captured with a net to obtain all relevant morphological data. A movie or video camera can also be used to obtain wing kinematics concurrently with airspeed measurements (see Section 3.2). The principal advantage of this method is that complete sets of biomechanical data can be obtained on the same insect in natural free flight. This technique has potentially wide application, not only for insects encountered in natural flight over bodies of water, but also for experiments in which insects are released in the middle of a lake. If such a released insect establishes a regular flight trajectory (constant orientation and speed), airspeeds and wing kinematics can then be determined.

## 3.2 Wing kinematics

A detailed description of wing motions can be obtained only through photographic means. For insects with high wingbeat frequencies (that is, 150 Hz or higher for many insects with asynchronous flight muscle), high-speed cinematography at 500 frames/sec or higher is essential to describe wing motions and configuration in detail. Lower speed cine-films (30–200 frames/sec) have also

**Figure 2.** Method for direct measurement of the airspeed of a flying insect. The insect is followed in a small motorboat such that the trajectories of the insect and boat are nominally parallel. Airspeed in the immediate vicinity of the insect is measured with an anemometer held laterally from the prow of the boat. Video or cine-films can also be taken to determine wing kinematics during natural free flight. The same insect can then be captured and appropriate morphological measurements made. (R. Dudley and R. B. Srygley following *Aphrissa statira* (Pieridae) in migratory flight over Lake Gatún, Republic of Panama. Photo courtesy of Carl Hansen.)

found wide application in studies of insect flight. Movie films provide high-resolution images for analysis, although the demands of lighting and focus usually require repeated camera adjustment during the filming process. More recently, the development of video technology has immensely simplified kinematic studies of insect flight. Video has considerable advantages over cine-film. Film development is unnecessary, video film is much cheaper than cine-film, and the success of filming can be evaluated immediately. The inherent disadvantage of video is that the effective filming frequency is fairly low (25 or 30 frames/sec). Spatial resolution is also much reduced relative to cine-film. However, the field portability and general user convenience of consumer video products suggest that this technology will receive widespread use in both laboratory and field studies of insect flight (see Chapter 3 for details of video recording and coordinate digitization of kinematic data). Also, high-speed video cameras (for example, NAC and Kodak), although expensive at present, are becoming increasingly available to biologists. When coupled with high-resolution monitors, such cameras may ultimately eliminate high-speed cinematography and its attendant difficulties from the experimental repertoire.

### 3.2.1 Two-dimensional analysis of wing kinematics

In normal hovering and forward flight, wing motions of a flying insect can be assumed to be symmetrical with respect to the longitudinal body axis. This assumption of symmetry during flight considerably simplifies the kinematic analysis, permitting geometric reconstruction of the three-dimensional wing-tip position from two-dimensional film images (see ref. 4 for mathematical details of the reconstruction procedure). Briefly, points of interest on the insect's wings and body are transformed from a coordinate system based on the film image to a coordinate system based on the insect's body. The three-dimensional position of the wing-tip relative to the body can then be described as a function of time through the wingbeat. In order to normalize all lengths measured from the film or video image, a reference length must be determined for each analysed flight sequence. If the filming speed is a factor of 30 or more higher than the wingbeat frequency, then the maximum projected wing length of the sequence can be used as the reference length (4). Alternatively, the projected body length can also be used as the reference length, provided that the longitudinal body axis is orthogonal to the optical axis of the camera.

The kinematic analysis of insect flight assumes that wing motions are confined to a stroke plane which forms an angle $\beta$ relative to horizontal (see *Figure 3*). This is an appropriate simplification for the wing motions of most insects studied thus far (in general, wing motions may also be assumed to follow simple harmonic motion). The maximum positional angle, $\phi_{max}$, and the minimum positional angle, $\phi_{min}$, refer to the wing-tip position in the stroke plane at the top and bottom of the half-stroke, respectively (*Figure 4*). Stroke amplitude $\Phi$, the angular extent of motion in the stroke plane, is given by $\phi_{max} - \phi_{min}$; the mean positional angle, $\bar{\phi}$, is equal to $(\phi_{max} + \phi_{min})/2$. Body angle $\chi$ is the angle relative to horizontal made by a line segment connecting the anterior and posterior tips of the body, when the insect is viewed laterally (see *Figure 3*).

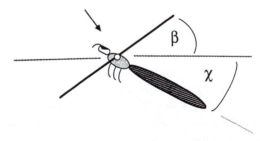

**Figure 3.** Body angle $\chi$ and stroke plane angle $\beta$ for a flying insect. Wing motions are assumed to be confined to a stroke plane oriented at angle $\beta$ relative to horizontal. The longitudinal axis of the body forms an angle $\chi$ with respect to horizontal, when the insect is viewed laterally. The arrow indicates the viewing perspective of *Figure 4*.

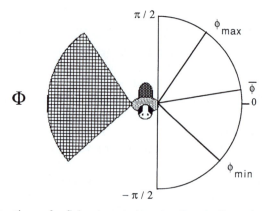

**Figure 4.** Wing motions of a flying insect viewed orthogonally to the stroke plane. The angular position $\phi$ describes the location of the wing tip in the stroke plane. The wing moves through an arc of amplitude $\Phi$ $(= \phi_{max} - \phi_{min})$ during a wingbeat. The mean positional angle of the wingbeat, $\bar{\phi}$, is equal to $(\phi_{max} + \phi_{min})/2$.

When using video films to determine the above kinematic parameters, the effective filming speed is often lower than the wingbeat frequency. In this case, it is necessary to properly identify the two video frames representing the top and bottom of the half-stroke. Because within any given wingbeat these frames do not in general immediately follow one another, analysis of composite wingbeats formed from non-sequential frames is necessary. In practice, one frame from several consecutive wingbeats is chosen that depicts the wing at the top of the half-stroke, and the video tape is then advanced frame-by-frame to a later frame depicting the wing at the bottom of the half-stroke. Multiple composite wingbeats should be analysed in any given flight sequence to determine average values for kinematic parameters. If wingbeat frequency greatly exceeds video filming frequency and discrete wing positions cannot be resolved, then the extreme positions of the wingstroke can be determined from the maximum and minimum extent of wing motion in the stroke plane.

If filming speeds exceed the wingbeat frequency by a factor of 30 or more, values of the latter quantity (typically represented by $n$) and the upstroke-to-downstroke ratio (US:DS) can be determined simply by counting the number of frames necessary to complete each half-stroke. Values of US:DS can also be determined from plots of the wing positional angle as a function of time, which procedure is more accurate when temporal interpolation between consecutive frames is necessary (4). If the wingbeat frequency is close to the filming speed, the number of video frames required to complete an integral number of wingbeats must be counted, and a mean value for the number of frames per wingbeat determined. Because of the low accuracy of this procedure, such an estimate should be repeated several times to derive an average wingbeat frequency for any given flight sequence. Acoustic recordings can also be used to measure wingbeat frequencies, using playback of the waveform on a storage oscilloscope or through a microcomputer-based data acquisition system to determine the fundamental frequency of the signal. Additionally, an optical tachometer has been devised that transduces the visual signal of oscillating wings into an analogue voltage output (9). This technique is particularly effective for use on small insects (<10 mg) that produce only low amplitude sounds during flight.

### 3.2.2 Three-dimensional kinematics

Determination of asymmetrical wing motions or complex flight paths requires methods supplemental to those described in Section 3.2.1. A three-dimensional description of wing and body motions can be obtained using two orthogonally arranged and synchronized cameras, or with techniques of stereophotogrammetry. Alternatively, a mirror can be positioned at 45 degrees to the image plane of one camera, allowing simultaneous filming of multiple views of a flying insect. A method for reconstructing three-dimensional trajectories from two cameras views has been developed with specific reference to animal flight [(10); see also ref. 11]. Infra-red illuminators used in conjunction with synchronized video cameras permit nocturnal tracking of insect trajectories in the field; three-dimensional reconstruction of such flight trajectories follows standard stereoscopic procedures (12).

From records of three-dimensional flight paths and manœuvres, a number of performance parameters such as manœuvrability and agility can be determined (see ref. 13 for parameter definitions). To date, however, no quantitative comparisons of turning performance have been made between different insect taxa. Turning radius is predicted theoretically to be proportional to the wing loading (Section 4.1.2), while agility (the maximum angular acceleration entering a turn) should be directly proportional to the wing span and inversely proportional to the moment of inertia of the wings and body about the longitudinal body axis. Confirmation of these predictions for flying insects would be highly desirable, as would be comparative studies of body orientations and rotations during manœuvres (see, for example, ref. 14).

### 3.3 Flow visualization

Flow visualization is a technique whereby tracer particles are used to make visible patterns of air movement around flying animals. Fields of neutrally buoyant helium bubbles have successfully been used to study the vortex wakes of bats (15) and birds (16). Patterns of air flow around tethered insects can be observed using lycopodium or talcum powder (17, 18), although to date the only free-flying insects thus examined have been butterflies during take-off (19, 20). The utility of flow visualization for insects is somewhat limited because of their small size and generally high wingbeat frequencies. However, certain large insects with low wingbeat frequencies (for example, butterflies, many moths, and some orthopterans) would be suitable candidates for wake visualization during free forward flight.

## 3.4 Force production of tethered and free-flying insects

By attaching tethered insects to force transducers (see Chapter 3 for technical details), the summed aerodynamic and inertial forces produced by the flapping wings can be measured. If kinematic and morphological data are also available, wing inertial forces can then be estimated and subtracted from total forces to obtain the net aerodynamic forces as a function of time (see refs. 21 and 22). Simultaneous electromyographic recordings permit correlations between wing position, muscle activity, and force output (23). An interesting experimental manipulation is to add weights to a free-flying animal to determine its maximum capacity for vertical force production (see ref. 24), and to evaluate the consequences of increased effective body mass for flight kinematics (see, for example, refs. 25 and 26). Care must be taken in such experiments to position the weights as near to the centre of body mass and as symmetrically about the longitudinal body axis as possible, in order to avoid creating additional rotational moments that may confound analysis of force output during flight. Potential demands of load-carrying and transient increases in body mass arise frequently in flying animals (for example, flight following food ingestion, transport of prey, flight of gravid females and of migrants with lipid reserves), but their possibly adverse effect on flight performance has not been systematically investigated.

# 4. Morphological and mechanical measurements

After a flying insect has been successfully filmed and its wing and body kinematics determined, morphological data from the same individual are necessary to implement aerodynamic analyses. This section introduces key morphological parameters used in the biomechanical analysis of insect flight (see ref. 27 for detailed discussion of theory and methods involved in these morphological measurements). Mechanical appraisal of isolated wings and bodies includes measurement of steady-state aerodynamic characteristics, structural testing of the wings, and functional morphology of the wing articulation and thorax.

## 4.1 Flight morphology

### 4.1.1 Wings

Knowledge of wing length $R$ (the distance from wing base to apex) and the total wing area $S$ (the area of both wing pairs) is essential for aerodynamic calculations (see *Figure 5*). The latter quantity can be measured directly from the wing planform (electronic leaf area meters used by botanists work well for large insect wings; see also Chapter 2 for area digitizing methods). The wing

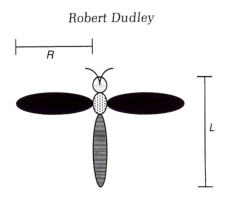

**Figure 5.** Dorsal view of an insect depicting the wing length $R$ and body length $L$. The area in black represents the total wing area $S$. Plan area ($S_{plan}$) is the projected area of the body in this dorsal perspective.

aspect ratio, AR ($= 4R^2/S$), is a useful overall descriptor of wing shape; higher aspect ratio indicates a more narrow wing. Wing mass $m_w$ includes the mass of both wing pairs (or both wings, for dipterans). Because the wing virtual mass $v$ (the mass of air accelerated by the wings) can be comparable in magnitude to the wing mass, virtual mass must also be calculated for purposes of bio-mechanical analysis, using the equations given in (27). Non-dimensional parameters that describe wing area and mass distribution along the wing span also figure prominently in the aerodynamic analysis of insect flight. Mathematical background and techniques necessary to determine the spanwise moments (and their non-dimensional radii) of wing area, wing mass, and wing virtual mass are discussed in ref. 27.

### 4.1.2 Bodies

Knowledge of the body mass $m$ (which includes $m_w$) is essential for analysis of the force balance of a flying insect. Because body mass can often change substantially during flights even of short duration, $m$ should be measured both before and after flight experiments. Use of the mean body mass is then appropriate for biomechanical analysis. Wing loading $p_w$ ($= mg/S$, where $g$ is gravitational acceleration) represents the average pressure imposed upon the air by the wings. Body length $L$ is conveniently non-dimensionalized by dividing by the wing length $R$ (see *Figure 5*). Calculation of a non-dimensional mean body diameter, $\hat{d}$ ($= (4m/\pi\rho_b L)^{1/2}$, where $\rho_b$ is the mean density of the body) is also useful for comparative purposes. Various non-dimensional parameters that describe the distribution of mass along the longitudinal body axis also figure in the mechanical analysis of body kinematics (27). For calculations of steady-state force coefficients (Section 4.2), a reference area for the body must be determined . Plan area ($S_{plan}$) is the projected area of the body as viewed dorsally, while frontal area ($S_{frontal}$) is the cross-sectional area of the body when viewed anteriorly.

## 4.2 Aerodynamic characteristics of wings and bodies

Force transducers can be used to measure drag (the aerodynamic force parallel to the relative airflow) and lift (the aerodynamic force perpendicular to flow) on isolated wings and bodies positioned in wind-tunnel airstreams. Transducer technology is discussed in Chapter 3; for application in flight biomechanics, strain-gauge and optoelectronic force transducers (both of which are based on deflection) are particularly useful. In such arrangements, the test object is mounted on a narrow, rigid attachment rod (known as a 'sting') that permits variable orientation of the object to the direction of airflow. The opposite end of the sting is connected to a transducer, the displacement of which is recorded electronically. Lift ($L$) and drag ($D$) forces are measured at varying angles of attack of the test object relative to on-coming airflow. Calibration of the apparatus is attained by applying known forces to the point on the sting at which the test object is attached. The small contribution of the sting to total aerodynamic forces is measured independently with the transducer and is then subtracted from values measured on test objects, which procedure assumes that interaction between the sting and the test object is negligible.

### 4.2.1 Force coefficients

Lift and drag data are reduced to non-dimensional force coefficients ($C_L$ and $C_D$, respectively) by means of the following formulae (see ref. 28):

$$C_L = 2*L/\rho SV^2, \tag{1}$$

$$C_D = 2*D/\rho SV^2, \tag{2}$$

where $\rho$ is the density of air, $S$ a reference area for the object, and $V$ the airspeed. For insect bodies, $S$ can be either the plan or frontal area of the body, while plan area is usually chosen as the reference area for wings. Values of $C_D$ and $C_L$ for insect bodies are advantageously plotted as a function of the Reynolds number $Re$ ($= lV/\nu$, where $l$ is some characteristic length of the test object, and $\nu$ is the kinematic viscosity of air; see also Chapter 11 for discussion of the $Re$). For wings, $C_L$ is typically given as a function of $C_D$ for any given airspeed (i.e. $Re$), with angle of attack of the wing plotted para-metrically. The characteristic length for wings is usually the mean wing chord ($= S/2R$), while $L$ is the characteristic length used for bodies. Characteristic lengths by definition do not vary with object orientation relative to airflow.

Although aerodynamic forces measured under static conditions can only partially resemble the dynamic phenomena that characterize insect flight, such measurements provide an important initial assessment of the magnitude of forces acting on moving wings and bodies. Use of static lift and drag coefficients is most appropriate when considering forces on the insect body during forward flight. Although airflow past the body is also influenced by the induced flow generated by the wings (Section 5.2.3), this latter component is in many cases small relative to the forward component of motion. Also, spatial and temporal variability in the induced flow field precludes precise characterization of the flow regime around the body, and use of the mean forward airspeed is an appropriate simplification. For flapping wings, however, flow magnitude varies even more dramatically both along the wing and through the course of a wingbeat. Effects of wing rotation, acceleration, and deformation (see ref. 29) also reduce the utility of steady-state force data. Steady-state lift and drag coefficients for insect wings should therefore not be overinterpreted. Even so, not only is knowledge of the maximum lift coefficient necessary in the mean coefficients method of aerodynamic analysis (Section 5.1), but wing drag coefficients derived under steady-state conditions of flow must be used to estimate the power requirements associated with wing drag forces (the profile power; Section 5.2.1), as the appropriate measurements of unsteady drag forces on wings have not been performed (see ref. 30).

### 4.2.2 Force balance in steady horizontal flight

In horizontal flight at a constant velocity, net vertical forces produced by the flapping wings, $F_{vert}$, must balance the body weight $mg$ less body lift, $L_b$, while a horizontal force or thrust ($F_{hor}$) must be generated to overcome the fluid drag ($D_b$) acting on the body (*Figure 6*). Although for many insects

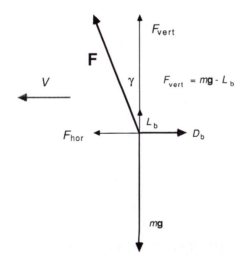

**Figure 6.** Force balance for an insect in horizontal flight at a constant airspeed $V$. Net vertical forces produced by the flapping wings, $F_{vert}$, must balance the body weight $mg$ less any body lift, $L_b$. Similarly, the net horizontal force, $F_{hor}$, must balance body drag, $D_b$. The magnitude of the resultant force vector **F** is given by $(F_{hor}^2 + F_{vert}^2)^{1/2}$, while the tilt of **F** with respect to vertical, $\gamma$, is given by $\tan^{-1}(F_{hor}/F_{vert})$.

thrust is small relative to vertical forces produced by the wings, the power required to overcome body drag (the so-called parasite power; Section 5.2.2) can be a substantial component of total power requirements at higher airspeeds. Body lift ($L_b$) is typically even smaller in magnitude than body drag, but at high airspeeds can contribute substantially to the overall force balance. Both body drag and lift can be measured directly on isolated bodies, or estimated using equations (1) and (2) provided that appropriate reference areas and force coefficients are available. Once $F_{hor}$ and $F_{vert}$ are known, the magnitude and orientation of the resultant force vector is easily calculated (see *Figure 6*).

## 4.3 Structural analyses of wings and the thoracic complex

By applying torque and point loads to wing surfaces and measuring the ensuing deformations, structural rigidity of wings and the implications of venational patterns can be evaluated. Mechanical instrumentation and tests appropriate for investigating the structural characteristics of insect wings are described in (31). Because flapping wings experience dynamic loads, the relationship of static measurements to forces on moving wings is not entirely clear. However, static tests coupled with estimates of aerodynamic and inertial torques during flight can provide a useful initial assessment of structural design (see refs. 32 and 33). Experimental manipulations, such as the cutting of specific veins or the excision of particular wing regions, are also possible in static mechanical testing.

The insect wing base, thorax, and associated musculature interact to form a complex three-dimensional structure, the functional details of which remain unresolved for most insect groups (34). Cinematography can be used to reveal the dynamics of thoracic and wing base deformation during tethered flight (see, for example, ref. 35), although again the similarity of such preparations to free flight performance is unknown. Hypothesized muscle actions based on thoracic dissections and existing descriptive morphology can also be evaluated electromyographically in tethered preparations (36). In general, however, wing kinematics can be highly variable both in steady forward flight and particularly during manœuvres. The complete range of functional versatility in wing motions and concomitant activity of the flight muscles has yet to be described for any insect.

# 5. Aerodynamic forces and power requirements

Given that kinematic and morphological data are available for a flying insect, aerodynamic forces and the associated mechanical power requirements of flight can be estimated. Analytical expressions for the aerodynamics and power requirements of hovering flight can be found in ref. 30, while current approaches to the aerodynamics of forward flight in animals are detailed in ref. 2.

## 5.1 Quasi-steady analysis

In the quasi-steady aerodynamic analysis of animal flight, the flapping wings are conceptually divided into a static series of spanwise sections operating at specific angles of attack and relative velocities (20). Aerodynamic consequences of such unsteady phenomena as wing rotation and acceleration are ignored, while instantaneous aerodynamic forces are assumed to be proportional to the square of the velocity of the wing section. In the mean coefficients method (37), a mean lift coefficient is determined such that net vertical forces equal to the body weight over the period of a wingbeat. Equations necessary to implement the mean coefficients method for forward flight can be found in refs. 38 and 39, and will be summarized here.

### 5.1.1 Relative velocity and aerodynamic forces

In forward flight, the relative air velocity $V_R$ of a wing section is determined from the vector sum of the forward airspeed of the insect, the induced velocity (Section 5.2.3), and the flapping velocity of the section. Instantaneous flapping velocities can be obtained from detailed analysis of wing kinematics (Section 3.2.1), although if simple harmonic motion of the wing can be assumed, precise information on wing position as a function of time is unnecessary. Aerodynamic forces acting on wing sections are assumed to be proportional to the square of the velocity component orthogonal to the longitudinal wing axis (that is, spanwise flow is ignored). Drag, the force parallel to the relative velocity (*Figure 7*), is estimated from equation (1) using an instantaneous profile drag coefficient ($C_{D,pro}$) for the wing section. Profile drag on the wing represents the summed effects of pressure and skin friction drag, while drag arising from the induced velocity is treated separately (Section 5.2.3). Values of $C_{d,pro}$ are determined from the effective angle of incidence of each wing section and the Reynolds number (29). Lift, the force perpendicular to the relative velocity, is then expressed using equation (2), although the lift coefficient, assumed to be a constant, remains unknown. The direction of lift forces must be evaluated from the effective angle of incidence of the section and the orientation of the relative velocity vector. Once lift and drag forces are derived, vertical components of both forces can be expressed for each wing section (see *Figure 7*).

### 5.1.2 Vertical force balance and the mean lift coefficient

By equating the body weight (less body lift) with summed vertical forces along the wing and through the wingbeat, a mean lift coefficient for the wings, $\overline{C}_L$, can be derived. If $\overline{C}_L$ exceeds maximum lift coefficients determined for airfoils under conditions of steady flow at similar Reynolds numbers, then

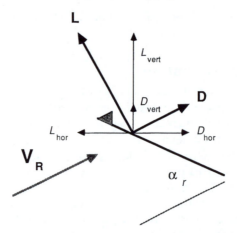

**Figure 7.** Aerodynamic forces on a wing section, the leading edge of which is indicated by a large triangle. The component of the relative velocity vector orthogonal to the longitudinal wing axis, $V_R$, is inclined at an effective angle of attack $\alpha_r$ to the wing. In quasi-steady analysis, lift ($L$) and drag ($D$) forces on the wing section are assumed to be proportional to the square of $V_R$. Lift and drag forces can be resolved into vertical ($L_{vert}$, $D_{vert}$) and horizontal ($L_{hor}$, $D_{hor}$) components. In the mean coefficients method, a mean lift coefficient ($\overline{C}_L$) is determined such that the summed vertical forces created by the flapping wings through the wingbeat equal the body weight (less any body lift).

unsteady aerodynamic phenomena are implicated. However, near agreement of $\overline{C}_L$ with empirically determined lift coefficients does not unequivocally disallow unsteady mechanisms, as instantaneous unsteady phenomena may still yield mean lift coefficients close to measured values (ref. 20; see also ref. 21). Given knowledge of $\overline{C}_L$, the relative contribution of vertical forces produced during the downstroke ($F_{vert,d}$) to total vertical forces can also be determined. Variation of $F_{vert,d}$ with forward airspeed is of particular interest, as in flying animals the downstroke generally assumes a greater role in offsetting body weight as airspeed increases. Thrust production is not evaluated in this analysis, because the small magnitude of body drag to weight results in a high sensitivity of horizontal force estimates to wing orientation and relative velocity (39).

## 5.2 Power requirements of flight

The total mechanical power required to fly can be evaluated by considering four individual components: profile power $P_{pro}$, parasite power $P_{par}$, induced power $P_{ind}$, and inertial power during the first half of a half-stroke, $P_{acc}$. Additionally, mechanical power may be required to maintain oscillations in the centre of body mass ($P_{cm}$). Mechanical power requirements are frequently expressed in mass-specific form (that is, W/kg body mass), and such usage is conventionally indicated by an asterisk (for example, $P^*_{acc}$).

### 5.2.1 Profile power

Profile power ($P_{pro}$) is the power required to overcome drag forces on the wings. At intermediate flight speeds, profile power for flapping wings has been treated as a constant fraction of the minimum power speed, the speed which maximizes flight time per unit energy (40). Profile power requirements in forward flight have also been evaluated assuming that upstroke forces are negligible (41). Neither of these methods is necessarily appropriate for analysis of flight in insects. An alternative approach calculates instantaneous quasi-steady values of wing drag and profile power, and derives an average profile power for the wingbeat (39). This method incorporates both spanwise and temporal variation in the relative velocity and profile drag coefficient ($C_{D,pro}$) of the wings. Profile power is determined from the product of the wing drag and the relative velocity:

$$P_{pro} = \rho \, C_{D,pro} \int_0^R c(r) \, V_R(r)^3 \, dr, \tag{3}$$

where $\rho$ is the mass density of air, $c(r)$ the wing chord at radial distance $r$ from the wing base, and $V_R(r)$ the relative velocity of the wing at distance $r$. Values of $C_{D,pro}$ include the effects of pressure drag and skin friction on the wing, but not the induced drag, which is treated separately (see Section 5.2.3). Profile power is evaluated instantaneously and summed for consecutive spanwise wing sections over the length of the wing $(0 - R)$, and an average profile power is then determined for the entire wingbeat. Equation (3) incorporates a factor of two [cf. equation (2)] to account for the profile power requirements of both wing pairs of a flying insect.

### 5.2.2 Parasite power

Parasite power $P_{par}$ is the power required to overcome drag forces on the body, and is equal to the product of the body drag and the forward airspeed:

$$P_{par} = D_b V. \tag{4}$$

Body drag is either measured directly (Section 4.2) or estimated using equation (2) and a body reference area. In the latter case, the body drag coefficient must be taken from the literature at the appropriate $Re$. As body drag is in many cases small relative to the weight of an insect, mechanical power estimates will not be overly sensitive to the choice of drag coefficient.

### 5.2.3 Induced power

Induced power ($P_{ind}$) is the power required to impart sufficient downwards momentum to the surrounding air so as to offset the body weight. In Rankine–Froude propeller theory, this mass flux is modelled by assuming that an actuator disk applies a constant pressure impulse to the air. The downwards airflow that ensues is assumed to be of constant magnitude (the induced velocity $V_i$), which in forward flight is determined from:

$$V_i = (mg - L_b)/2\rho A_0 (V^2 + V_i^2)^{1/2}, \tag{5}$$

where $A_0$ is the area of the actuator disk (see refs. 8 and 39) for discussion of appropriate disk areas in forward flight). Induced power is then given by the product of the body weight (less any lift forces; Section 4.2) and the induced velocity:

$$P_{ind} = (mg - L_b) V_i. \tag{6}$$

A more detailed means of determining the induced velocity and power in hovering flight incorporates spatial and temporal fluctuations in the pressure impulse applied by the actuator disk (29). A vortex model to determine induced velocity and power has also been developed for forward flapping flight (41), but, as mentioned previously, the upstroke is assumed not to generate useful aerodynamic forces.

The sum of profile, parasite, and induced power requirements is termed the aerodynamic power, $P_{aero}$. In hovering flight, $P_{aero}$ is simply the sum of $P_{pro}$ and $P_{ind}$, as body drag and hence parasite power are negligible. The aerodynamic efficiency of flight, $\eta_a$, equals the ratio of $P_{ind}$ to $P_{aero}$, and indicates the relative efficiency of the flapping wings in generating a downwards momentum flux to offset the body weight.

### 5.2.4 Inertial power

In flapping flight, inertial power is required to accelerate the wing mass and virtual mass. Inertial power during the first half of a half-stroke, $P_{acc}$, is given by:

$$P_{acc} = 2 \, nI(\mathrm{d}\phi/\mathrm{d}t)^2_{max}, \tag{7}$$

where $I$ is the moment of inertia of the wing mass and virtual mass, and $(\mathrm{d}\phi/\mathrm{d}t)_{max}$ is the maximum angular velocity attained by the wing during the half-stroke (30). Values of $(\mathrm{d}\phi/\mathrm{d}t)_{max}$ can either be determined from actual kinematic data or estimated by assuming simple harmonic motion of the wings. If the downstroke and upstroke are of unequal duration, maximum angular velocities must be determined separately for each half-stroke. The moment of inertia for the wing mass and virtual mass can be determined using the formula given in ref. 30 (see also Chapter 4).

If the kinetic energy of the oscillating wing mass and virtual mass can be stored as elastic strain energy in thoracic structures and later released, then inertial power requirements through the wingbeat will be zero. Mechanical power requirements for this case of perfect elastic energy storage ($P_{per}$) simply equal the aerodynamic power, $P_{aero}$. Alternatively, if there is no elastic energy storage of wing inertial energy, then supplementary power will be required to accelerate the wing during the first half of a half-stroke. During the second half of the half-stroke, however, the negative power requirements that characterize wing deceleration are close to zero, as the metabolic costs of negative work (tension generation during stretching) are much reduced relative to energetic expenditures associated with positive work (30). Aerodynamic power requirements over the same period, which in many cases are small relative to the inertial power, can be supplied by the kinetic energy of the decelerating wings. The power output as averaged over the half-stroke will then equal one-half the sum of the aerodynamic power requirements and the inertial power during the first half of the half-stroke (30). Thus, total mechanical power requirements given zero elastic energy storage, $P_{zero}$, are given by:

$$P_{zero} = (P_{aero} + P_{acc})/2. \qquad (8)$$

Depending on the extent of elastic energy storage, actual power requirements lie between $P_{per}$ and $P_{zero}$, which represent the lower and upper bounds, respectively, of energetic expenditure during flight.

### 5.2.5 Power to accelerate and lift the centre of body mass

Most flying vertebrates demonstrate only minimal oscillations of the centre of body mass, and the concomitant power required to maintain such fluctuations in kinetic and potential energy is negligible. Many insects, however, display erratic movements and sudden vertical and lateral accelerations in otherwise approximately rectilinear flight trajectories. Such behaviour may entail substantial energetic expenditures. For example, erratic flight path oscillations of palatable Neotropical butterflies increased total mechanical power requirements by an average of 43% (42).

To estimate the power ($P_{cm}$) associated with flight path oscillations, the position of the centre of body mass as a function of time must be derived photographically. Fluctuations in vertical and horizontal kinetic energy and in potential energy are then determined in a manner analogous to the analysis for cursorial locomotion (see Chapter 4). The sum of positive increments in the total energy of the centre of mass is then divided by the period of analysed flight to obtain the average power required to maintain these fluctuations. For flying animals, $P_{cm}$ is added to either $P_{per}$ or $P_{zero}$ to obtain total mechanical power requirements, as the kinetic energy of oscillating wings cannot supplement the mechanical energy of an animal suspended in air.

### 5.3 The summed power curve

When individual components of mechanical power for a flying animal are known at different airspeeds (see, for example, *Figure 8a*), the total power requirements for the two cases of zero ($P_{zero}$) and perfect ($P_{per}$) elastic energy storage can be calculated. The ensuing 'power curve' (*Figure 8b*) relates the mechanical costs of flight to forward airspeed (40). Power curves have found widespread ecological application in that the maximum range speed (that is,

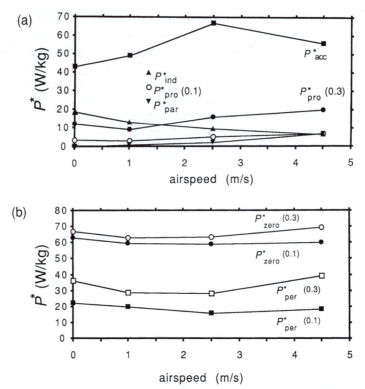

**Figure 8.** Mechanical power requirements for a bumblebee worker at different airspeeds (modified from ref. 39). (a) Individual components of the power curve. ■:$P^*_{acc}$, inertial power during the first half of a half-stroke; ●:$P^*_{pro}$ (0.3), profile power assuming a mean profile drag coefficient of 0.3; ○:$P^*_{pro}$ (0.1), profile power given a mean profile drag coefficient of 0.1; ▲:$P^*_{ind}$, induced power; ▼:$P^*_{par}$, parasite power. (b) Total mechanical power requirements as a function of forward airspeed. ○:$P^*_{zero}$ (0.3), total power assuming zero elastic energy storage and a mean profile drag coefficient of 0.3. ●:$P^*_{zero}$ (0.1), total power assuming zero elastic energy storage and a mean profile drag coefficient of 0.1. □:$P^*_{per}$ (0.3), total power assuming perfect elastic energy storage and a mean profile drag coefficient of 0.3. ■:$P^*_{per}$ (0.1), total power assuming perfect elastic energy storage and a mean profile drag coefficient of 0.1.

minimum energy expenditure per unit distance) and the minimum power speed can easily be determined (see ref. 1).

## 5.4 Metabolic power input and flight muscle efficiency

Mechanical power is related to metabolic power input by the flight muscle efficiency $\eta_m$, which indicates the conversion efficiency of chemical to mechanical energy. Flight muscle efficiency may vary as a function of flight speed, particularly if wing kinematics (for example, stroke amplitude and wingbeat frequency) change systematically with forward airspeed. Energetics of hovering flight have frequently been measured in insects (see refs. 43 and 44 for reviews), and estimates of $\eta_m$ during hovering range from 5–29%, depending on the extent of elastic energy storage (30). Using a sophisticated closed circuit wind tunnel, metabolic rates of bumble-bees in free forward flight were shown to be fairly constant over the range of hovering up to an airspeed of 4 m/sec (45). This constancy paralleled the relatively invariant mechanical power requirements estimated for bumble-bees over the same speed range (39). In bumble-bees, the flapping velocity of the wings is high relative to the forward airspeed over the range considered (low advance ratio; see ref. 4). Changes in forward airspeed thus have little effect on the wing relative velocity and profile power requirements [see equation (3)], and thus total power requirements do not vary dramatically. In such cases, then, hovering metabolism may be similar to that in forward flight. Values of $\eta_m$ can thus be derived by comparing energetic data for hovering with mechanical power estimates for free forward flight. Flight metabolism can also be measured on insects tethered in the jet of a closed circuit wind tunnel [see ref. 46; but the same limitations of tethering as described previously (Section 2.1) also apply in this case].

# 6. Using flight biomechanics to study insect diversity

Functional consequences of the morphological diversity that characterizes the pterygote Insecta are virtually unexplored. Almost no data are available for the flight characteristics of most insect taxa. For example, beetles are commonly thought to be poor fliers, and most large beetles do indeed appear awkward in flight. However, the majority of beetle species are less than 10 mm in body length (47), and are often fast, highly manœuvrable, and capable of hovering. Given the taxonomic diversity and numerical preponderance of insects in most terrestrial ecosystems, comparative surveys of flight kinematics and morphology would be of considerable utility and consequence.

One important evolutionary tendency of the pterygote Insecta is the multiple independent origin of asynchronous flight muscle, a muscle type characterized by high rates of contraction and hence high wingbeat frequencies (48). Because of the dependence of aerodynamic forces upon the square of wing velocity [equations (1) and (2)], an increased wingbeat frequency permits a disproportionate decrease in wing surface area if equivalent forces are to be generated. Wing reduction or radical modification of one wing pair is common in asynchronous insect fliers. For example, beetle forewings have become modified to act as protective armour (the elytra), while the hindwings of flies have become miniaturized to act as gyroscopic organs (the halteres). Possession of asynchronous flight muscle characterizes three of the four largest insect orders and over 75% of extant pterygote insect species (49). The

117

aerodynamic consequences of high wingbeat frequencies thus have figured prominently in insect evolution. Mechanistic aspects of this process can only be understood through comparative investigation of flight kinematics, aerodynamics, and morphology for both synchronous and asynchronous insect fliers.

Biomechanical analysis of flight permits such varied aspects of performance as maximum flight speed, turning radius, and capacity for acceleration to be determined. What is the linkage between such quantitative results and the real world as viewed through the compound eyes of an insect? How fast do insects really fly in nature? What implications do the power requirements of flight (and the underlying morphological and aerodynamic considerations) have for flight behaviour and temporal partitioning of activity? How frequently are the inherent capacities for manœuvrability and acceleration actually used? The functional elucidation of biological mechanisms is a useful and challenging intellectual process. However, only when placed in the context of organismal performance in natural environments will the full utility of biomechanical analysis be realized.

# 7. Appendix: abbreviations

| | |
|---|---|
| $A_0$ | Area of actuator disk |
| AR | Wing aspect ratio ($= 4R^2/S$) |
| $c(r)$ | Wing chord at radial distance $r$ from wing base |
| $C_D$ | Drag coefficient |
| $C_{D,\text{pro}}$ | Profile drag coefficient |
| $C_L$ | Lift coefficient |
| $\overline{C_L}$ | Mean lift coefficient |
| $\hat{d}$ | Non-dimensional mean body diameter |
| $D$ | Aerodynamic drag |
| $D_b$ | Body drag |
| $F_{\text{hor}}$ | Average horizontal forces produced by insect in steady forward flight ($= D_b$) |
| $F_{\text{vert}}$ | Average vertical forces produced in steady forward flight ($= mg - L_b$) |
| $F_{\text{vert,d}}$ | Vertical force production during the downstroke |
| **g** | Gravitational acceleration |
| $I$ | Moment of inertia of the wing mass and wing virtual mass |
| $l$ | Characteristic length of a wing or body |
| $L$ | Body length (Section 4.1.2), or aerodynamic lift (Section 4.2) |
| $L_b$ | Body lift |
| $m$ | Body mass |
| $m_w$ | Wing mass |
| $n$ | Wingbeat frequency |
| $p_w$ | Wing loading ($= m\mathbf{g}/S$) |

| | |
|---|---|
| $P_{acc}$ | Inertial power requirements during the first half of a half-stroke |
| $P_{aero}$ | Aerodynamic power requirements ($= P_{par} + P_{ind} + P_{pro}$) |
| $P_{cm}$ | Mechanical power necessary to lift and accelerate the centre of body mass |
| $P_{ind}$ | Induced power requirements |
| $P_{par}$ | Parasite power requirements |
| $P_{per}$ | Total mechanical power requirements given perfect elastic energy storage |
| $P_{pro}$ | Profile power requirements |
| $P_{zero}$ | Total mechanical power requirements given zero elastic energy storage |
| $r$ | Radial position along wing |
| $R$ | Wing length |
| $Re$ | Reynolds number |
| $S$ | Wing area (Section 4.1.1), or reference area for force coefficients (Section 4.2.1) |
| $S_{plan}$ | Plan area |
| $S_{frontal}$ | Frontal area |
| US:DS | Ratio of upstroke to downstroke |
| $v$ | Wing virtual mass |
| $V$ | Airspeed |
| $V_i$ | Induced velocity |
| $V_R$ | Relative velocity |
| | |
| $\beta$ | Stroke plane angle |
| $\eta_a$ | Aerodynamic efficiency ($= P_{ind}/P_{aero}$) |
| $\eta_m$ | Muscle efficiency |
| $\nu$ | Kinematic viscosity of air |
| $\rho$ | Mass density of air |
| $\rho_b$ | Body mass density |
| $\bar{\phi}$ | Mean positional angle ($= (\phi_{max} + \phi_{min})/2$) |
| $\phi_{max}$ | Maximum positional angle of the wing |
| $\phi_{min}$ | Minimum positional angle of the wing |
| $(d\phi/dt)_{max}$ | Maximum angular velocity attained by the wing during a half-stroke |
| $\Phi$ | Stroke amplitude ($= \phi_{max} - \phi_{min}$) |
| $\chi$ | Body angle in flight |

# Acknowledgement

The author thanks Min Lu for kindly drawing the illustrations.

# References

1. Pennycuick, C. J. (1989). *Bird flight performance*. Oxford University Press, New York.
2. Norberg, U. M. (1990). *Vertebrate flight*. Springer-Verlag, Berlin.
3. Miyan, J. A. and Ewing, A. W. (1985). *Phil. Trans. R. Soc. Lond.* B, **311**, 271.
4. Ellington, C. P. (1984). *Phil. Trans. R. Soc. Lond.* B, **305**, 41.
5. Ennos, A. R. (1989). *J. exp. Biol.*, **142**, 49.
6. Dudley, R. and Ellington, C. P. (1990). *J. exp. Biol.*, **148**, 19.
7. DeVries, P. J. and Dudley, R. (1990). *Physiol. Zool.*, **63**, 235.
8. Dudley, R. and DeVries, P. J. (1990). *J. comp. Physiol.* A, **167**, 145.
9. Unwin, D. M. and Ellington, C. P. (1979). *J. exp. Biol.*, **82**, 377.
10. Rayner, J. M. V. and Aldridge, H. D. J. N. (1985). *J. exp. Biol.*, **118**, 247.
11. Dahmen, H. J. and Zeil, J. (1984). *Proc. R. Soc. Lond.* B, **222**, 107.
12. Riley, J. R., Smith, A. D., and Bettany, B. W. (1990). *Physiol. Ent.*, **15**, 73.
13. Norberg, U. M. and Rayner, J. M. V. (1987). *Phil. Trans. R. Soc. Lond.* B, **316**, 335.
14. Wagner, H. (1986). *Phil. Trans. R. Soc. Lond.* B, **312**, 527.
15. Rayner, J. M. V., Jones, G., and Thomas, A. (1986). *Nature, Lond.*, **321**, 162.
16. Spedding, G. R., Rayner, J. M. V., and Pennycuick, C. J. (1984). *J. exp. Biol.*, **111**, 81.
17. Brodskii, A. K. and Ivanov, V. D. (1984). *Zool. Zh.*, **63**, 197.
18. Brodskii, A. K. and Ivanov, V. D. (1985). *Priroda*, **10**, 74.
19. Ellington, C. P. (1980). Vortices and hovering flight. In *Instationäre Effekte an Schwingenden Tierflügeln* (ed. W. Nachtigall), pp. 64–101. Franz Steiner, Wiesbaden.
20. Ellington, C. P. (1984). *Phil. Trans. R. Soc. Lond.* B, **305**, 1.
21. Cloupeau, M., Devillers, J. F., and Devezeaux, D. (1979). *J. exp. Biol.*, **80**, 1.
22. Wilkin, P. J. (1990). *J. Kans. Ent. Soc.*, **63**, 316.
23. Esch, H., Nachtigall, W., and Kogge, S. N. (1975). *J. comp. Physiol.*, **100**, 147.
24. Marden, J. H. (1987). *J. exp. Biol.*, **130**, 235.
25. Videler, J. J., Groenewegen, A., Gnodde, M., and Vossebelt, G. (1988). *J. exp. Biol.*, **134**, 185.
26. Pennycuick, C. J., Fuller, M. R., and McAllister, L. (1989). *J. exp. Biol.*, **142**, 17.
27. Ellington, C. P. (1984). *Phil. Trans. R. Soc. Lond.* B, **305**, 17.
28. Vogel, S. (1981). *Life in moving fluids*. Willard Grant Press, Boston.
29. Ellington, C. P. (1984). *Phil. Trans. R. Soc. Lond.* B, **305**, 79.
30. Ellington, C. P. (1984). *Phil. Trans. R. Soc. Lond.* B, **305**, 145.
31. Hepburn, H. R. and Chandler, H. D. (1980). Materials testing of arthropod cuticle preparations. In *Cuticle techniques in arthropods* (ed. T. A. Miller), pp. 1–44. Springer-Verlag, Berlin.
32. Wootton, R. J. (1981). *J. Zool., Lond.*, **193**, 447.
33. Ennos, A. R. (1988). *J. exp. Biol.*, **140**, 137.
34. Kammer, A. E. (1985). Flying. In *Comprehensive insect physiology, biochemistry, and physiology* (ed. G. A. Kerkut and L. I. Gilbert), Vol. 5, pp. 491–552. Pergamon Press, Oxford.
35. Miyan, J. A. and Ewing, A. W. (1988). *J. exp. Biol.*, **136**, 229.

36. Rheuben, M. B. and Kammer, A. E. (1987). *J. exp. Biol.*, **131**, 373.
37. Osborne, M. F. M. (1951). *J. exp. Biol.*, **28**, 221.
38. Norberg, U. M. (1976). *J. exp. Biol.*, **65**, 459.
39. Dudley, R. and Ellington, C. P. (1990). *J. exp. Biol.*, **148**, 53.
40. Pennycuick, C. J. (1975). Mechanics of flight. In *Avian biology* (ed. D. S. Farner and J. R. King), Vol. 5, pp. 1–75. Academic Press, London.
41. Rayner, J. M. V. (1979). *J. Fluid Mech.*, **91**, 731.
42. Dudley, R. (1991). *J. exp. Biol.* **159**, 335.
43. Kammer, A. E. and Heinrich, B. (1978). *Adv. Insect Physiol.*, **13**, 133.
44. Casey, T. M. (1989). Oxygen consumption during flight. In *Insect flight* (ed. G. J. Goldsworthy and C. H. Wheeler), pp. 257–72. CRC Press, Boca Raton, Florida.
45. Ellington, C. P., Machin, K. E., and Casey, T. M. (1990). *Nature, Lond.*, **347**, 472.
46. Nachtigall, W., Rothe, U., Feller, P., and Jungmann, R. (1989). *J. Comp. Physiol.* B, **158**, 729.
47. May, R. M. (1978). The dynamics and diversity of insect faunas. In *Diversity of insect faunas* (ed. L. A. Mound and N. Waloff), pp. 188–204. Blackwell Scientific Publications, Oxford.
48. Pringle, J. W. S. (1981). *J. exp. Biol.*, **94**, 1.
49. Dudley, R. (1991). Comparative biomechanics and the evolutionary diversification of flying insect morphology. In *The unity of evolutionary biology: Proceedings of the Fourth International Congress of Systematic and Evolutionary Biology* (ed. E. Dudley), pp. 503–14. Dioscorides Press, Portland, Oregon.

# 6

# In vivo measurement of bone strain and tendon force

ANDREW A. BIEWENER

## 1. Introduction

An understanding of the design and mechanical performance of biological structures ultimately depends on measurements of the forces applied to them and the resulting deformations developed within them during their use. For structural elements of the limbs of vertebrate species, this may involve *in vivo* measurement of muscle forces, tendon forces and bone strains during various functional activities, such as locomotion. While these *in vivo* measurements can provide detailed and precise data, their invasive nature introduces the possibility of diminished performance of the animal being studied. This possibility must be weighed carefully against the use of non-invasive approaches, such as those based on ground reaction force recordings in relation to limb kinematics (see Chapter 3). If carefully planned and executed, however, experimental *in vivo* recordings of bone strain and muscle–tendon force generally provide data of high quality and value.

*In vivo* bone strain measurements have been used to interpret patterns of bone loading during locomotion (1–3), as well as during feeding (4). In addition, *in vitro* measurements of bone strain made under simulated loading conditions can also help to elucidate the functional significance of skeletal structures that may not be amenable to *in vivo* recordings. Measurements of muscle force and tendon strain provide important data for understanding not only the mechanical requirements of muscle force generation over an animal's range of locomotor activity (5), but the role of the nervous system in controlling the activity of muscles to perform various motor tasks (6, 7). In addition, these measurements can be used to assess an animal's ability to conserve energy by means of elastic strain energy storage and recovery during locomotion (5), which depends on tendon geometry in relation to tendon stiffness (8, 9). Finally, quantitative measurements of *in vivo* bone strain are essential for interpreting adaptive tissue remodelling processes in response to changes in functional strain pattern within a bone (10–13). In this context, *in vivo* bone strain data may be used to verify finite element models of skeletal structures by comparing calculated strain distributions with measured strains at specific sites on the surface of the bone (see Chapter 7). Alternatively, *in vivo* bone strain data, together with tissue mechanical properties, may be used to define better the loading conditions of the bone model under investigation.

## 2. *In vivo* measurement of surface bone strain

### 2.1 Strain-gauges

The measurement of surface bone strains by strain-gauges bonded directly to the bone's surface during functional activity in live animals was first made possible with the advent of rapid, self-catalysing cyanoacrylate adhesives (14). Strain-gauges are variable resistive elements that change electrical resistance proportional to the localized deformation (strain) of the underlying material to which the gauge is bonded. It is important to recognize, therefore, that strain is only transduced at specific sites on the bone. The most commonly used strain-gauges in biological applications are constantan metal foil gauges, having either 120 or 350 ohms resistance. Changes in resistance are due to changes in both the cross-sectional area of the foil grid (compressive strain increases the foil's cross-sectional area, lowering its resistance) and the specific resistivity of the foil (compression also lowers specific resistivity; tension produces the opposite effects, increasing gauge resistance). Strain is transduced only along the primary axis ($y$, *Figure 1*) of the gauge in units of microstrain ($\mu\varepsilon$, or strain $\times 10^{-6}$). Semiconductor strain-gauges, which have greater sensitivity compared to metal foil resistance strain-gauges and are useful in the fabrication of force transducers, are not recommended for these kinds of measurements.

Constantan foil gauges are commonly mounted on either a thin polyimide or polyester resin backing, that is tough but flexible (the polyester resin backing is the more flexible and, we find, the easier to work with of the two), giving the gauges a long fatigue life (prolonged tolerance to cyclical loading). Other backings are available for industrial applications, but these are generally unnecessary or inappropriate for biological applications. Leading manufacturers of strain-gauges are Tokyo Sokki Kenkyujo Ltd., Measurements Group, Inc., and BLH Electronics, Inc. In general, the cost of gauges from the former is roughly 50% less than that of the latter two suppliers (based on a comparison of 1990 prices). Strain-gauges come in a variety of shapes and sizes, their suitability depending on the specific application and the size of the bone that is being studied.

Both single element and rectangular stacked rosette metal foil strain-gauges have generally been used to record surface bone strains *in vivo* (*Figure 1*). Rectangular rosette strain-gauges comprise three independent strain-gauge elements stacked directly on one another, with two gauge elements ($\varepsilon_a$ and $\varepsilon_c$) aligned at 90 degrees to each other and the third element ($\varepsilon_b$)

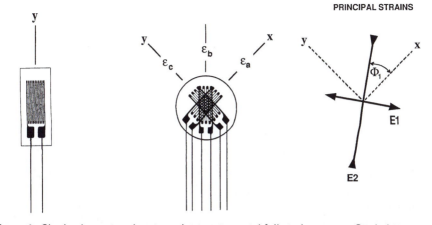

**Figure 1.** Single element and rectangular rosette metal foil strain-gauges. Strain is transduced along the *y*-axis of the single element strain-gauge. The rosette gauge comprises three independent gauge elements ($\varepsilon_a$, $\varepsilon_b$, and $\varepsilon_c$) stacked on one another, from which the maximum (tensile, *E1*) and minimum (compressive, *E2*) principal strains can be determined. These are depicted to the right. The orientation ($\Phi_1$) of the maximum principal strain (*E1*) calculated by equation (3) is referenced to the $\varepsilon_a$ axis (defined here as the *x*-axis).

bisecting these at an angle of 45 degrees. An alternative configuration is the delta rosette, in which the three gauge elements are positioned at 60 degrees to one another in a triangular configuration (15). This configuration is less well suited to *in vivo* applications, however, because of the greater difficulty of connecting lead wires to the gauge elements and in bundling the wires together as they are led away from the gauge recording site (see Section 2.3).

## 2.2 Measurement of principal strains and their orientation

Rosette strain-gauges enable determination of the maximum and minimum principal strains (see Chapter 1) and their orientations to the $\varepsilon_a$ axis of the strain-gauge *irrespective* of the gauge's orientation on the bone's surface. This is a key advantage for most biological applications, because the primary axis of strain is almost always unknown to the investigator beforehand. The maximum (*E1*) and minimum (*E2*) principal strains are calculated from the strains measured along the three axes of the gauge according to the following equations (which assume a planar state of strain)

$$E1 = (\varepsilon_a + \varepsilon_c)/2 + [(\varepsilon_a - \varepsilon_c)^2 + (2\varepsilon_b - \varepsilon_a - \varepsilon_c)^2]^{1/2}/2 \qquad (1)$$

$$E2 = (\varepsilon_a + \varepsilon_c)/2 - [(\varepsilon_a - \varepsilon_c)^2 + (2\varepsilon_b - \varepsilon_a - \varepsilon_c)^2]^{1/2}/2 \qquad (2)$$

The angle ($\Phi_1$) of the maximum principal strain (*E1*) to the $\varepsilon_a$ gauge is calculated as

$$2\Phi_1 = \tan^{-1}\left[(2\varepsilon_b - \varepsilon_a - \varepsilon_c)/(\varepsilon_a - \varepsilon_c)\right] \tag{3}$$

The principal axes can be identified with respect to the $\varepsilon_a$ axis by testing for the following conditions (15):

$$0 < \Phi_1 < 90° \quad \text{when } \varepsilon_b > 1/2 \,(\varepsilon_a + \varepsilon_c)$$
$$-90° < \Phi_1 < 0 \text{ when } \varepsilon_b < 1/2 \,(\varepsilon_a + \varepsilon_c)$$
$$\Phi_1 = 0 \qquad \text{when } \varepsilon_b > \varepsilon_c \text{ and } \varepsilon_a = E1$$
$$\Phi_1 = 90° \qquad \text{when } \varepsilon_a < \varepsilon_c \text{ and } \varepsilon_a = E2$$

The principal strain axes can be referenced, in turn, to a structural axis of interest (such as the bone's longitudinal axis) by knowing the orientation of the $\varepsilon_a$ gauge to that axis. This can be measured directly following attachment of the gauge to the bone's surface or from a radiograph of the bone, by using a radio-opaque marker (such as small piece of solder) epoxied to the gauge's top surface that clearly marks the axis of the $\varepsilon_a$ gauge on the bone.

In contrast to rosette strain-gauges, single element strain-gauges transduce strain only along the primary axis (*y*) of the gauge (*Figure 1*). Hence, the strain measured by the gauge will vary depending on the gauge's orientation on the bone. Single element strain gauges are satisfactory either (i) when the principal axes of strain are known or (ii) when strain is only of interest along one specified axis. In most cases, however, the principal axes of strain at a given site on a bone are unknown beforehand, necessitating the use of rosette strain gauges to determine the magnitude and orientation of principal strain.

## 2.3 Strain-gauge lead wire soldering and insulation

Strain-gauges generally can be purchased with or without pre-soldered un-insulated wire leads from the grid tabs of each gauge. Generally, the un-insulated wire leads facilitate soldering *insulated* lead wires to the six tabs of a rosette strain-gauge, but are necessary for soldering to a single element gauge. An advantage of the strain gauges supplied by Tokyo Sokki Kenkyujo is that the lead wires soldered to the gauge tabs are circular in cross-section, making them easy to bend in any direction compared to the leads of the other suppliers, which are flat and more difficult to manipulate. Etched Teflon insulated 36-gauge lead wire (Measurements Group, Inc.) minimizes tissue reactivity and provides a durable subcutaneous connection from the gauge elements to an externalized connector. For larger animals, or when longer lead wires are needed, the resistance of small gauge lead wires may be significant, requiring the use of thicker wire (see Section 4.1 below). Different-coloured pairs of lead wires are critically important to identifying individual elements when rosette gauges are used. When recordings from more than one rosette strain-gauge are to be made, the lead wires of the additional gauge(s)

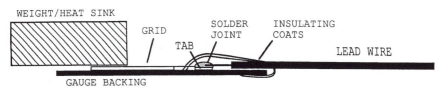

**Figure 2**. Drawing showing the position of the strain-gauge leads over the gauge backing. Etched Teflon insulated leads are soldered either directly to the gauge tabs or to un-insulated wire leads connecting to the tabs of the gauge. Insulating layers of polyurethane M-coat 'A' and xylene-based acrylic M-coat 'D' (Measurements Group, Inc.) are applied over and beneath the lead wires to ensure a tight seal between the strain-gauge backing and the wires.

should be colour-coated near their ends with an indelible marker or finger nail polish to distinguish clearly which set of six leads belongs to which gauge. This is of particular importance when the lead wires are to be soldered to an external plug following attachment of the gauges to the bone of the animal.

One of the most frequent sources of strain-gauge failure involves fatigue or insulation breakdown of the lead wires at their solder joints. To minimize the likelihood of this, it is extremely important that well-soldered connections of minimal size are made as close to the tabs of the gauge as possible (*over* the gauge backing, see *Figure 2*). The following are recommended for use in the following protocol for lead wire soldering:

- dissecting microscope
- resin core solder and flux
- low wattage soldering iron with a fine tip
- watchmaker's forceps

---

**Protocol 1.** Lead wire soldering

1. Coat 0.5 mm of the exposed end of the insulated lead wire with flux and tin.[a]

2. Affix the strain gauge and lead wires securely in position by taping down to a note card. The lead wires should be positioned side by side, with the tinned ends overlying the gauge tabs.

3. Apply flux to both the tinned lead wire and the gauge tab. In the case of soldering to a pre-soldered lead wire (as for a rosette-gauge), wrap the lead wire over and around the tinned insulated lead wire tip before applying flux.

4. After soldering, clean all lead wire connections thoroughly with 100% alcohol using a cotton swab.[b]

**Protocol 1.** *Continued*

5. Check each gauge for an intact circuit and appropriate resistance before insulating the lead wires. Ensure that none of the solder connections are 'cold' by lightly tugging at each.

6. Pot the gauge solder connections for insulation and mechanical support.

[a] Tinning simply involves coating the exposed tip of the lead wire with solder prior to soldering the lead wire to the gauge tab.

[b] MEK is frequently recommended by manufacturers, but being a much stronger solvent, it can damage the integrity of the foil gauge and its backing.

---

Because excessive heat can easily disrupt the foil gauge from its backing, a small steel block may be placed over the end of the strain-gauge to help dissipate heat.

Several different coatings are available for insulation and support of the exposed lead wire solder connections to the gauge. We use separate coatings of a polyurethane (M-coat 'A') and a xylene-based latex (M-coat 'D', Measurements Group, Inc.) which provide reliable insulation of reasonable strength, while retaining good flexibility. Several thin applications of each coating, together with small solder connections, help limit the overall bulk of the gauge, which may prove critical for successful positioning and attachment on the bone. These coatings require several (5–7) days for full cure before use. Other oven-cure epoxies may also be used; however, their slower curing time can prove a nuisance, as epoxy may flow to unwanted regions (such as the undersurface of the gauge). If possible, it is important to avoid air bubbles in the coatings and to ensure that the undersurface of the lead wires are well insulated where they pass over the gauge backing. Finally, for large animals, a lead wire strain relief assembly may be constructed by passing the lead wires through a moulded epoxy flange that can be screwed down to the bone at a short distance from the gauge (see Section 2.5.1 below). In smaller animals, however, this is not feasible, and the lead wires must be secured by tying them down with suture to adjacent fascia. A short polyethylene sleeve (5 mm cut from tubing of appropriate size) epoxied around the lead wires can also be used to help anchor the lead wires with the suture (silk or a synthetic braided nylon suture is recommended).

## 2.4 Recording sites

Because strain gauges record strain only from the region directly beneath that to which they are bonded, selection of strain recording sites on a bone should be given careful thought. When the strain at a specific site on the bone is to be determined, it is important to judge whether this site is surgically accessible and will not cause undue injury to surrounding soft tissues. As space is often limited for attachment of a strain-gauge to a bone, particularly in smaller animals, small gauges are generally preferred for *in vivo* strain recordings. Such space or size limitations, in fact, may necessitate the use of a single element versus a rosette strain-gauge, or preclude the measurement of bone strain altogether. For instance, we (10) recorded bone strains on the tibiotarsus of 4-week-old chicks using single element gauges (grid dimensions: 1.5 × 2 mm, type FLE-11, Tokyo Sokki Kenkyujo) because of the small size of the bone at this age (about 5 mm midshaft diameter). Rosette strain-gauges were used to verify principal strains and their orientations at functionally equivalent sites on the bone at older ages.

If the goal of the recordings is to infer patterns of whole bone loading, it is critical that the gauges not be unduly influenced by stress concentrations produced by localized pull of muscles, tendons or ligaments. Localized strain engendered by muscles or tendons attaching to a bone, however, is always a potential problem. In general, the midshaft cortices of a long bone are those which are the most free of muscle attachments. Midshaft bone strain recordings, therefore, can provide a reliable assessment of strains (and forces) transmitted through the shaft of the bone, which reflect the net distribution of loads applied to the bone. An additional consideration is that the maximum strains experienced by a long bone are most likely to occur at the midshaft due to bending moments produced by eccentric (off-axis) loading at the bone's ends. To assess the relative contributions of axial versus bending loads, strain-gauges must be positioned on opposing cortices in the primary plane of bending (for example, anterior and posterior cortices, if bending is in the antero-posterior plane). However, the plane of bending is often unknown or may shift during the gait cycle or at different speeds and gaits. This possibility can be resolved by attaching three gauges about the perimeter of the bone, which allows the distribution of normal strains to be mapped out across the entire cortex.

### 2.4.1 Determination of normal strain and the neutral axis of a cortical bone cross-section

The distribution of strains *normal* to the bone's cross-sectional plane can be determined based on the measurement of strain at three sites around the circumference of a bone's shaft at a given level (13, 16). The approach assumes a uniform distribution of normal strain in the section and only requires the use of single element strain-gauges oriented normal to the plane of the bone's cross-section (that is, along the longitudinal axis of the bone). Recordings from one or more rosette strain-gauges to assess the principal strains and their orientation may also be used to determine the normal strains at these locations (see Section 2.8). By measuring the normal strains at three defined locations around the bone's circumference, a system of three linear equations consisting of three unknowns ($a$, $b$, and $c$) can be solved to establish the equation describing the planar distribution of normal strains across the bone's cortex. Before this can be done, the cross-sectional shape of the bone's

## PLANAR DISTRIBUTION OF NORMAL STRAINS

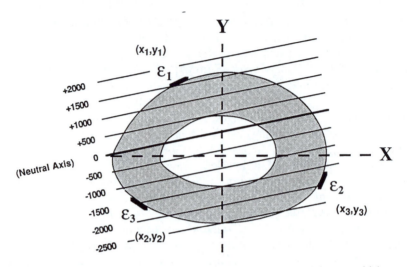

**Figure 3.** A drawing of a bone cross-section showing the three sites at which normal strains are recorded to allow determination of the equation describing the distribution of normal strains in the plane of section. The heavy solid line depicts the neutral axis ($\varepsilon = 0$), which must be solved first before establishing the equations for the other strain isoclines. See text for additional details.

cortex must be digitized (see Chapter 2) and the bone's shape and the position of the strain-gauges defined with respect to a reference coordinate system.

Let $\varepsilon_1$, $\varepsilon_2$, and $\varepsilon_3$ be the normal strains recorded at sites 1, 2, and 3 on the bone's surface, having coordinates $(x_1, y_1)$, $(x_2, y_2)$ and $(x_3, y_3)$, respectively (*Figure 3*). The following equations can be written:

$$\varepsilon_1 = ax_1 + by_1 + c \tag{4}$$

$$\varepsilon_2 = ax_2 + by_2 + c \tag{5}$$

$$\varepsilon_3 = ax_3 + by_3 + c. \tag{6}$$

The equation for the line defining the neutral axis ($\varepsilon = 0$) can then be written

$$0 = ax + by + c$$
$$y = -a/b\ x - c/b \tag{7}$$

allowing the linear equations that define other strain 'isoclines' (for example, $-500\ \mu\varepsilon$, $+500\ \mu\varepsilon$, $-1000\ \mu\varepsilon$, $+1000\ \mu\varepsilon$, etc.) to be determined. By doing so, the gradient of normal strain can be mapped out across the bone's cortex at any given time during the loading cycle.

## 2.5 Bonding strain-gauges to living bone

### 2.5.1 Bone recording sites

The sites at which bone strains are to be recorded from should be given careful thought before proceeding to surgery. A thorough knowledge of the anatomical organization of the bone and soft tissues surrounding and attaching to it is essential. Dissection of frozen or preserved material prior to surgery is greatly recommended. It is generally advisable that no major muscle, tendinous or ligamentous origin or insertion exist at or near to where the strain-gauges are to be attached. This is important not only because disruption of these tissues will increase the likelihood of the animal being lame, but under most circumstances the objective is to measure bone strains that are relevant to loads transmitted by the *whole* bone, not those transmitted by the localized pull of a muscle or tendon. For most long bones, the midshaft cortices of the bone are those which are most free from significant muscle or ligamentous attachments. Hence, the midshaft of a long bone provides an ideal region for recording *in vivo* bone strains. This is particularly important if an assessment of the distribution of strain across the bone's entire cortex is desired, which requires that three recording sites be identified at a given cross-sectional level of the bone. For consistency and to reduce inter-individual variability in the strain data that are collected, it is also important that the same recording sites are used in all animals. Anatomical landmarks and metric measurements can be used to help minimize variation resulting from this source.

### 2.5.2 Surgical procedures

Surgical exposure and attachment of strain-gauges to the surface of a living animal's bone requires appropriate anaesthesia and sterile surgical procedure. It is assumed that anyone attempting to use this technique will have had the necessary surgical training or, if not, the professional assistance of a veterinarian, and will adhere to currently accepted guide-lines for the proper care and analgesia of the species being studied. A methyl-cyanoacrylate adhesive is required to attach the strain-gauge to the cleaned surface on the bone. This adhesive can be purchased from a strain-gauge manufacturer or can be obtained at the hardware store under the genre of 'super glues' at a considerably lower cost. Presumably, the quality control of the latter is also less than that of the more expensive specialist's variety. In our experience, however, use of fresh glue purchased from a local hardware store has proved reliable.

*Protocol 2* outlines the basic procedure for preparing an attachment site on the bone and bonding the strain-gauge to it in a sterile manner. In addition to the standard surgical supplies and instruments, the following supplies and instruments are recommended:

- periosteal elevator
- retractors

- MEK or ether
- sterile polyethylene examining glove
- sterile cotton swabs
- methyl-cyanoacrylate adhesive
- digital ohm meter
- bridge amplifier and oscilloscope

---

**Protocol 2.** Surgical attachment of strain-gauges to living bone

  1. Sterilize the strain-gauges and lead wires using a gas sterilizer. If a gas sterilizer is unavailable, the gauges may be soaked in an antibacterial solution for a few hours as a reasonably safe alternative. If this is done, the gauges and lead wires should be rinsed with sterile saline early in the surgical procedure and allowed to dry before being used.

  2. Retract overlying musculature and tendons/ligaments (a second pair of hands is needed here).

  3. Using a scalpel, cut a small rectangular 'window' out from the periosteum (about twice the area of the gauge) and scrape free from the bone with a periosteal elevator.

  4. Clean and dry the underlying mineralized bone surface by means of a series of applications of ether or MEK, using a cotton-tipped applicator and light scraping of the bone surface with the periosteal elevator.

  5. Stop any bleeding at the bone's surface, as well as from the adjacent periosteum and musculature before bonding the gauge. When dried and cleaned, the surface of the bone should have a dull white appearance. This is crucial for successful bonding of the gauge.

  6. Place a small drop of cyanoacrylate adhesive on the undersurface of the gauge (this should be done by a 'non-scrubbed' assistant to the procedure, such as the anaesthesiologist), and press the gauge down firmly in position on the bone. A finger can be used if there is sufficient space; otherwise, a small flat stainless steel bar surrounded by a polyethylene sleeve made from tubing can be used to exert pressure. Sterile polyethylene examining gloves (cut off at the fingers) work well as a lining placed over the finger of the surgical glove to keep the investigator's glove tip from being glued to the gauge.

  7. Maintain stable, firm pressure on the gauge for at least 60 sec.

  8. Keep the lead wires stationary for another 3 to 5 min to allow for more complete polymerization of the adhesive.

  9. Tie down the lead wires with suture (silk or braided nylon) to nearby fascia at multiple sites for strain relief.[a]

10. Pass the lead wires subcutaneously to a connector positioned at some distance from the gauge attachment site (this may be directly out to the surface of the limb, if the animal is large enough and willing to carry the connectors taped to its leg). A thin length of stainless steel tubing (ID 4 to 5 mm) can be used to help feed the lead wires through the fascia under the skin. A large size sterile knitting needle also works well for this purpose.

11. Colour-coat gauge leads to code their identity and location on the bone.

12. Check gauge resistances for intact circuits.[b]

13. Suture close all wounds and solder the lead wires to the connector.

[a] For lead wire strain relief in larger animals, a small epoxy flange may be used to secure the lead wires rigidly to the bone with a stainless steel screw (12, 14). This requires first drilling and then tapping a hole in the bone's cortex to mount the screw. Although this procedure is more involved, it provides considerably better strain relief for the lead wires.

[b] Connecting the gauge circuits to the bridge amplifier and displaying the strain-gauge signals on an oscilloscope is strongly recommended following attachment of all the strain-gauges but prior to closing the animal's wounds. This provides a much better test of the quality of the strain-gauge's adhesion to the bone and the electronic circuits than simply verifying each gauge's resistance. If a strain-gauge is determined to be faulty at this time (high-frequency spiking on an otherwise stable baseline is a common indication of faulty insulation or a loose attachment), it is a fairly easy matter to remove the gauge from the bone, clean the bone surface, and mount a second gauge. This is certainly preferable to finding out that the gauge is faulty *after* the animal has recovered from surgery and anaesthesia.

---

For large species (for example, horse, goat, and dog), standard 6-pin connectors work well for each 3-element rosette strain-gauge attached to the bone. For smaller animals, customized miniconnectors can be made inexpensively from integrated circuit wire-wrap socket pins and receptacles.

## 2.6 Recording and A/D sampling of strain data

### 2.6.1 Bridge amplifier and wiring considerations

Lead wires from each strain-gauge element are connected via a shielded cable to a Wheatstone bridge amplifier circuit (for example, Vishay model 2200, Measurements Group, Inc.). The bridge circuit converts resistance changes of the gauge into an amplified voltage output. As variable resistive elements, each strain-gauge forms an 'active' arm of a Wheatstone bridge (in what is commonly referred to as a quarter-bridge configuration). Metal foil strain-gauges are manufactured to have a gauge factor ($S_g$) of about 2, which specifies their sensitivity measured as a resistance change for a given magnitude of strain ($S_g \varepsilon = \Delta R/R$). For example, a strain of 0.001 ($+1000 \mu\varepsilon$) will produce a 0.25-ohm (0.21%) increase in the resistance of a 120-ohm metal foil strain-gauge having a gauge factor of 2.1. The sensitivity of these strain-gauges can be considered uniform over the full range of temperatures experienced by biological tissues (their recommended operating range typically is from $-30$ to $80\,^\circ$C). General performance specifications for each set of strain gauges purchased are provided by the manufacturer.

In their final position attached to a bone (or to a transducer, see below), strain-gauges rarely exhibit their nominal resistance (120 or 350 ohms), but vary from this by a few tenths of an ohm. This results from deformation of the strain-gauge during its attachment to the bone and/or the presence of lead wire resistance. Because the strain-gauge's resistance now differs from that of the precision internal 'dummy' resistor located within the bridge amplifier (which is matched to the strain-gauge's manufactured resistance), the resistance difference produces a baseline or DC offset in the signal voltage. To zero the voltage output of the bridge circuit to correspond to a state of zero strain, the strain-gauge (SG) is 'balanced' against the opposing arms of the bridge (*Figure 4A*). Zero output is achieved when $R_{SG}/R_2 = R_3'/R_4'$, where $R_3'$ is the effective parallel resistance of $R_3 + R_{BAL}$. Bridge amplifiers typically have a balance range of $\pm 1000\,\mu\varepsilon$, or greater. A state of zero load may be established, for example, when the animal is at rest and the instrumented limb is held off the ground in a relaxed state.

This is also the best time to record calibrations (volts/microstrain) of each strain-gauge signal. Most commercially available bridge amplifier/signal conditioners come with a calibration switch that can be toggled to produce a voltage output corresponding to a known tensile or compressive strain ($+1000\,\mu\varepsilon$ or $-1000\,\mu\varepsilon$ in the case of the Vishay 2200). This is achieved by means of precision 'shunt' calibration resistors that are placed in parallel with opposing arms of the bridge ($R_3$ or $R_4$) when the calibration switch is thrown, producing a resistance change equivalent to $\pm 1000\,\mu\varepsilon$. In general, it is *not* a good idea to shunt the active (strain-gauge) arm of the bridge, because possible errors may be introduced by lead wire desensitization (see below). Shunt calibration establishes the integrity and gain of the internal bridge amplifier circuit itself, not the ability of the strain-gauge to *measure* strain. The latter depends on the quality of soldering and insulation, minimizing lead wire resistance and most importantly, the quality of the gauge's adhesion to the underlying surface of the bone (or other material). It is a good idea to calibrate all strain channels at the end of each recording session to ensure that no changes in sensitivity have occurred. This also provides an additional check on the integrity of each strain-gauge circuit. The strain calibrations should be stored on computer disk or recording tape (see Section 2.7), together with each day's experimental strain recordings.

### i. Lead wire desensitization

If the length of lead wires is great enough, especially for narrow gauge wire (over 1 metre for 36-gauge wire), resistance of the lead wires, themselves, can be a problem. Lead wire resistance causes desensitization of the strain-gauge and limits the balance range of the gauge. If sufficiently large, lead wire resistance may prevent achieving a zero balance of the bridge amplifier

**A.**

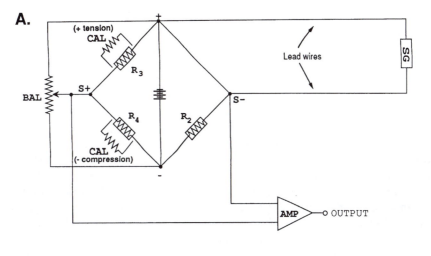

**B.**

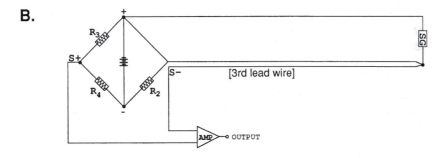

**Figure 4.** Quarter-bridge configuration used for individual strain-gauge channels. The implanted strain-gauge (SG) forms the 'active' arm of the bridge. A variable balance resistor (BAL) located in parallel with the opposing arms is used to achieve a zero voltage output when the animal is at rest and the bone is unloaded. By means of a toggle switch, a shunt calibration resistor (CAL) can be placed in parallel with one of the opposing arms of the bridge to simulate either $+1000\,\mu\varepsilon$ ($R_3$) or $-1000\,\mu\varepsilon$ ($R_4$).

circuit. To minimize the likelihood of this, thicker lead wires should be used in connecting the strain-gauge leads on the animal to the bridge amplifier. An assortment of multi-lead shielded cable is available from standard electronics suppliers. A shielded cable having six lead wires is the minimal number required for most applications (involving the use of two rosette strain-gauges). The cable shielding should be connected to the ground pin of each bridge amplifier input plug but left unconnected at the animal.

If lead wire resistance is a problem, as when recordings are made of the animal while it runs (or flies) overground, requiring the use of very long cables, a third lead wire must be soldered to one tab of each gauge to balance the lead wire resistance of the strain gauge (the 'active' arm) at the adjacent bridge resistor (*Figure 4B*). Use of a third lead wire, however, leaves less space for packaging the lead wires, increases the overall bulk of the strain gauge and increases the risk of lead wire failure.

An alternative approach that works satisfactorily in most situations is to use a resistor equal to the resistance of one lead wire, jumpered to the appropriate pins of the bridge amplifier input plug. The resistance per unit length of the lead wire can easily be determined using a digital ohmeter. This approach, however, does not compensate for differences in the temperature of the external resistor relative to that of the lead wire passing to the animal, which can cause drift in the signal and altered sensitivity. Under most biological conditions, these thermal effects are quite small in comparison to the magnitude of strain recorded at the surface of a bone and in comparison to the variation in strains typically recorded among individual animals. Temperature (or other noise induced) shifts in the strain voltage signal that may arise by using a two-wire quarter bridge configuration are chiefly the concern of material scientists and engineers concerned with more precise measurements of strain (that is, in the range of $\pm 20\,\mu\varepsilon$).

In addition to the potential effect of temperature on lead wire resistance, temperature changes of the strain-gauges, themselves, can also cause changes in their resistance and drift in the signal voltage. While this may be a concern for *in vitro* or materials testing applications, once again this is not usually a problem for species that maintain a stable body temperature. For ectothermic species, changes in body temperature over longer experimental periods may be a concern. To minimize the possibility of temperature-related drifts in the signal, it is always a good idea to keep the length of lead wires between the animal and the amplifier as short as possible and free from convection. Finally, whereas bridge excitation (the excitation voltage applied to each Wheatstone bridge) is also a concern for most industrial applications due to considerations of power dissipation in relation to heat gain of the strain-gauge, the high thermal conductivity of living tissues generally makes this a trivial concern in most biological applications. In our experience, a bridge excitation of 2.0 volts results in stable signals from implanted strain-gauges obtained over several hours of recording.

## ii. Noise

In general, noise is not a serious concern in recording *in vivo* surface bone strains. The strain signals obtained are typically quite 'clean' (i.e. they have a high signal-to-noise ratio); this is in contrast to the signals obtained when muscles are depolarized (see Chapter 8), which are of low magnitude at the source and must be amplified considerably before being recorded. In addition, because of their low noise component, *in vivo* strain signals usually do not require filtering (see *Figure 6*).

## 2.7 Sampling, data storage, and analysis considerations

Strain voltage signals and their calibrations either can be sampled directly into a computer via a 12-bit A/D converter (for example, an IBM-AT or 386 computer equipped with a DASH-16F, Metrabyte Corp., or DT2801A, DataTranslation) or stored on FM magnetic recording tape and played back into a computer at a later time. Parallel output of the signals to a storage oscilloscope is also useful for checking calibration levels, verifying that all circuits/strain-gauges are intact, and for general monitoring purposes during an experiment. Sampling at 200 Hz for a 1 sec cycle (stride) period is sufficient for most studies to provide detailed and accurate representation of the basic time-course of strain. To analyse higher frequency components of the signal, however, higher sampling rates are required. The sampling frequency should be minimally twice the highest frequency of interest to avoid aliasing of the sampled signal (see Chapter 3). Once acquired and stored on the computer, the principal strains and their orientations can be computed from the raw strains of a rosette strain-gauge using equations (1), (2), and (3). A BASIC program (PSCV.BAS) is included that performs these calculations.

## 2.8 Transformation of principal strains to principal stresses

Carter (17) outlines a procedure by which the strains and stresses along the principal structural axes of the bone (longitudinal and transverse axes), as well as the principal stresses and their orientation, can be determined from principal strain data. The approach considers bone to be a transversely orthotropic material, in which the bone's relevant material properties are defined along its longitudinal and transverse axes. The derivation given here generally follows that presented by Carter, who gives additional details that the reader may wish to consult. Strains in the principal material directions ($\varepsilon_1$, $\varepsilon_2$, $\gamma_{12}$; in which $\gamma_{12}$ is the shear strain; see *Figure 5*) are calculated from the principal strains (*E1* and *E2*) based on the following strain transformation equation

$$\begin{bmatrix} \varepsilon_1 \\ \varepsilon_2 \\ \gamma_{12}/2 \end{bmatrix} = \mathbf{T} \begin{bmatrix} E1 \\ E2 \\ 0 \end{bmatrix} \tag{8}$$

where $\mathbf{T}$ is the transformation matrix for strains in the two principal material directions (longitudinal: $\varepsilon_1$, and transverse: $\varepsilon_2$)

$$\mathbf{T} = \begin{bmatrix} \cos^2 \Phi & \sin^2 \Phi & -2 \sin \Phi \cos \Phi \\ \sin^2 \Phi & \cos^2 \Phi & 2 \sin \Phi \cos \Phi \\ \sin \Phi \cos \Phi & -\sin \Phi \cos \Phi & \cos^2 \Phi - \sin^2 \Phi \end{bmatrix} \tag{9}$$

and $\Phi$ is the orientation of the maximum principal strain (*E1*) to the bone's longitudinal axis. The stresses in the principal material directions ($\sigma_1$, $\sigma_2$, $\tau_{12}$;

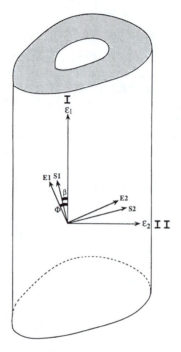

**Figure 5.** Principal strains (*E1* and *E2*) and principal stresses (*S1* and *S2*) defined with respect to the principal material directions of the bone (longitudinal: I and transverse: II). The maximum principal strain (*E1*) is oriented at an angle (Φ) and the maximum principal stress at angle (β) to the bone's longitudinal axis. Because the bone is anisotropic, principal stresses are not aligned with the principal strains. (Adapted from ref. 17.)

in which $\tau_{12}$ is the shear stress) are then determined from the strains ($\varepsilon_1$, $\varepsilon_2$, $\gamma_{12}$) by

$$\begin{bmatrix} \sigma_1 \\ \sigma_2 \\ \tau_{12} \end{bmatrix} = \mathbf{S} \begin{bmatrix} \varepsilon_1 \\ \varepsilon_2 \\ \gamma_{12}/2 \end{bmatrix} \tag{10}$$

where **S** is the reduced stiffness matrix. Values of the bone's transverse elastic modulus and shear modulus are assigned in proportion to the longitudinal elastic modulus ($\mathbf{E_{11}}$) based on empirical values of the Poisson's ratio for compact cortical bone [see Carter (17) for details]. Comparative data for the material properties of bone can be found in Currey (18), so that these relations can be applied to bones with differing elastic properties. Based on empirical data for the above material properties, the reduced stiffness matrix can be defined with respect to the longitudinal elastic modulus as

$$\mathbf{S} = \mathbf{E_{11}} \begin{bmatrix} 1.16 & 0.36 & 0 \\ 0.36 & 0.79 & 0 \\ 0 & 0 & 0.38 \end{bmatrix} \tag{11}$$

**138**

Yielding,

$$
\begin{bmatrix} \sigma_1 \\ \sigma_2 \\ \tau_{12} \end{bmatrix} = E_{11} \begin{bmatrix} 1.16 & 0.36 & 0 \\ 0.36 & 0.79 & 0 \\ 0 & 0 & 0.19 \end{bmatrix} \begin{bmatrix} \varepsilon_1 \\ \varepsilon_2 \\ \gamma_{12} \end{bmatrix} \tag{12}
$$

The magnitudes of the principal stresses ($S1$ and $S2$) and their orientation with respect to the longitudinal axis of the bone ($\beta$) are then calculated from

$$
S1, S2 = 0.5 \, (\sigma_1 + \sigma_2) \pm [0.25 \, (\sigma_1 - \sigma_2)^2 + \tau_{12}^2]^{0.5} \tag{13}
$$

and

$$
\beta = 0.5 \tan^{-1} [2\tau_{12}/(\sigma_1 - \sigma_2)]. \tag{14}
$$

Finally, it is also possible to calculate the net transverse (bending) and axial components of *force* applied to the bone (or other biological structure) once the longitudinal stress data have been obtained from *in vitro* strain data (2), based on measurements of the bone's cross-sectional geometry (see Chapter 2) and standard formulae for the flexure of beams (see Chapter 1).

## 2.9 Biological examples and their significance

A critical aspect and a potential concern of *in vivo* bone strain recordings is their localized assessment of strain; specifically, to what extent can a localized strain measurement be used to infer 'whole-bone' loading? In a study of bone strain in the radius and tibia of goats (1), we found that when the region of the bone is relatively free from localized muscle attachment, such as at the diaphysis of a long bone, strain patterns are consistent with those that would be expected from stresses transmitted through the whole of the bone (*Figure 6*). At sites close to muscular, tendinous, or ligamentous attachments to the bone, however, localized stress concentrations are likely to influence the strains measured by the gauge.

In our study of goats, we also showed that peak strains in these two bones were similar to those measured in the same two long bones of dogs (3) at equivalent points of gait (for example, trot-gallop transition or maximum gallop) but at different absolute speeds. Dogs trot and gallop at faster speeds than goats, but experience similar peak strains at these higher speeds by having disproportionately larger bones (increased cortical area and second moment of area). Peak strains of $-2000$ to $-2500 \, \mu\varepsilon$ were measured in the two species at a fast gallop, which is roughly one-fourth of the failure strain ($-8500 \, \mu\varepsilon$, or 0.0085) of bone tissue. Similar peak strains also occur in the long bones of horses at their fastest trotting speed (2, 3).

In addition to considerations of peak strain magnitude and safety factor, the *in vivo* strain recordings show that the majority (80–90%) of strain experienced by these long bones results from bending, rather than axial compression. The relative components of strain due to bending versus

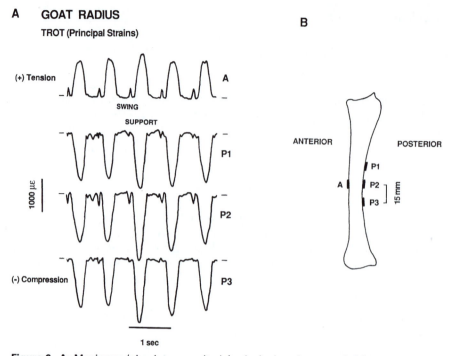

**Figure 6. A**: Maximum (absolute magnitude) principal strains recorded from rectangular rosette gauges attached to the midshaft of a goat radius during five strides of a trot. Absolute maximum principal strain is tensile (*E1*) on the anterior (A) cortex and compressive (*E2*) at three sites (*P1, P2, P3*) on the posterior cortex. These recording sites are shown in **B**. Dash marks indicate zero strain levels determined by balancing the gauge channels when the animal's limb was held in an elevated and relaxed position prior to the recording session. Bone strain rises to a maximum midway through support and falls to near zero during the swing phase of the stride. Principal strains recorded from the posterior gauges (*P1, P2,* and *P3*) are quite similar in pattern and magnitude, showing that whole bone loading patterns can be inferred from localized recordings of surface bone strain at the midshaft of long bones, if free of nearby muscle and tendon attachments. The decrease in strain magnitude from *P1* to *P3* is consistent with the decrease in bending moment expected. Tension recorded in the anterior cortex is 75% of the magnitude of compression in the posterior cortex, indicating that the bone is loaded mainly in bending, which is superimposed on axial compression. (Adapted from ref. 1.)

compression can be easily determined by comparing the magnitudes of strain recorded on opposing cortices of the bone at a given cross-sectional level. As shown in the midshaft of the goat radius (*Figure 6*), the anterior cortex experiences tensile strain ($\varepsilon_t = E1 = +1200\,\mu\varepsilon$), whereas the posterior cortex is loaded in compression ($\varepsilon_c = E2 = -1600\,\mu\varepsilon$ at a moderate trot). Strain due to bending, therefore, is

$$\varepsilon_b = \pm0.5\,(|\varepsilon_t| + |\varepsilon_c|) \tag{15}$$

**140**

$(= \pm 1400\,\mu\varepsilon)$ and axial compressive strain is

$$\varepsilon_a = \varepsilon_t - \varepsilon_b \tag{16}$$

$(= -200\,\mu\varepsilon)$. Recordings of strain on opposing midshaft cortices assume that the gauges are aligned with the primary plane of bending. If this is not the case, or given an eccentric compressive axis (with respect to the bone's longitudinal axis), the peak magnitude and the distribution of strain calculated within the bone's cortex will be in error. In this case, three strain-gauges attached to different locations about the bone's periphery are required to determine the plane of bending at a given cross-sectional level throughout the loading cycle (see Section 2.4.1). Although this approach can identify changes in the plane of bending during the gait cycle, once again if single element gauges are used to record normal strains, information about the principal strains and their orientation cannot be obtained.

# 3. *In vivo* measurement of muscle force

Measurements of muscle force *in vivo* can be accomplished by attaching a force transducer to a muscle's tendon and monitoring the force transmitted by the tendon. Consequently, this approach is limited to muscles having reasonably long (>1 cm) tendons. For a more general assessment of muscle forces exerted during locomotion, an indirect approach based on force platform and kinematic recordings of the limb may be preferred (see Chapter 3).

## 3.1 Transducer design

A common design of tendon force transducers is in the shape of an 'E', forming a 'buckle' that can be slipped on to the tendon (*Figure 7*). These transducers are machined from stainless steel to fit the size of the tendon. The quality of the 'fit' is probably the most critical aspect of a successful design. After machining to form the basic shape and size of the transducer, hand filing and abrasive polishing with progressively finer grit emery paper is required (finishing with 600 grit paper). Digitally automated laser machining or photochemical etching and electropolishing (19) can be employed to produce a transducer having high surface smoothness and precise, miniature dimensions. Tension within the tendon is transduced by a small strain-gauge mounted on the central 'arm' of these transducers. When the muscle contracts, tension in the tendon bends the transducer's arm, which is monitored by the strain-gauge (as in the case of the spring blade transducing arms of the force platforms described in Chapter 3). Consequently, use of these transducers requires *in situ* calibration of the transducer on the tendon to determine its voltage output in relation to force.

**Protocol 3.** Force-buckle fabrication

1. Machine stainless-steel buckle to appropriate dimensions.

2. Polish all surfaces with progressively finer grit emery (carborundum) paper.

3. Clamp single-element semiconductor or metal foil gauge in place on transducer arm using an oven-cured epoxy.[a] Semiconductor gauges are considerably more fragile, but have a much higher gauge factor (typically 50) and so, give a larger signal output.

4. Place in oven to cure (approximately 60 °C for 4 h).

5. Solder etched Teflon (36-gauge) lead wires to the gauge tabs and clean solder joints with 100% alcohol.

6. Insulate with additional epoxy (return to oven for curing).

7. Coat buckle transducer assembly with Parylene C,[b] a biologically inert compound that minimizes tissue inflammation and fibrous tissue formation around the transducer after implantation. For shorter-term recordings (1 to 3 days), a less durable silicon rubber coating (Dow Corning) may be used.

8. Calibrate transducer using a length of appropriate sized twine or cord.[c]

[a] We use AE-15 (Measurements Group, Inc.). An oven-cured epoxy should be used to ensure a high-quality bond for long-term attachment of the strain-gauge to the transducer. Large spring paper-clips work well for clamping the strain gauge in place during the adhesive's cure.
[b] Parylene C is applied as a vapour under vacuum to form an excellent non-tissue reactive, thin insulating protective layer over the stainless steel arms, strain-gauge, and lead wires of the transducer. This process is best carried out in an industrial setting (Viking Technology, Inc., or Para Tech Coating Co.).
[c] It is critical that the transducer be calibrated externally on appropriate sized twine or cord *before and after* its use in the animal. This is necessary to ensure that no change in the transducer's sensitivity has occurred (particularly when the experiments are to be carried out over a long period of time—weeks to months). Even so, the transducer must still be calibrated *in situ* on the tendon to determine *in vivo* force levels (see below).

## 3.2 Surgical procedures

Attachment of the buckle transducer to the tendon of interest most often is a straightforward surgical procedure (once again using proper surgical technique and anaesthesia). In the case of an agonist muscle group, such as the triceps surae of mammals, separate transducers may be required on individual tendons to monitor the recruitment of force among individual muscles (7). This is particularly important in studies of the motor control of locomotion, in which the distribution of recruitment among agonist muscles for force generation at a particular joint is a central question. This is also important when two muscles have differing contractile properties or differing mechanical advantages at the joint. On the other hand, when all muscles and their tendons have a

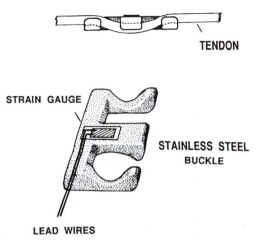

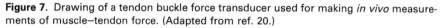

**Figure 7.** Drawing of a tendon buckle force transducer used for making *in vivo* measurements of muscle–tendon force. (Adapted from ref. 20.)

similar mechanical advantage, or when the tendons are too small to allow attachment of individual transducers, a single transducer attached to a group of tendons (such as the Achilles tendon) may be preferred (20). In some cases (as when separate recordings are to be made on individual tendons that are grouped together), the tendinous sheath must be cut open to slip the buckle transducer on to the exposed tendon(s). Preservation of the tendon's sheath is generally impossible, in any event, due to wear between the buckle and the tendon.

For accurate and reliable measurement of force the transducer must fit snugly on the tendon, but not so tight as to cause ischaemia. Because many tendons change cross-sectional area along their length, slippage of the transducer can alter its sensitivity during the experiment. An advantage of fibrous tissue that may form around the transducer is that this generally prevents slippage after a few days. If the fibrous tissue response is excessive, however, this can also lead to an unwanted change in transducer sensitivity. On the other hand, an overly tight fit can cause abrasion and damage to the tendon, as well as taking up slack that is normally present in the tendon and which may adversely affect the muscle's force–length properties. All of these possibilities are likely to diminish the quality of the animal's performance and hence, the value of the data that are collected.

## 3.3 Wiring and recording procedures

Recordings of muscle–tendon force from these strain-gauge-based transducers employs the use of a Wheatstone bridge amplifier as outlined above. The output can be displayed on an oscilloscope or a chart recorder, or it can be digitally sampled and entered into a computer for later analysis. Similar considerations apply to recording and sampling as outlined above for bone strain recordings. The lead wires from the transducer should be secured to adjacent fascia with suture and can be run subcutaneously to an externalized connector. For small animals, such as rodents, a connector mounted on the skull provides the best approach for attaching a recording cable to the animal. The connector should be anchored to the skull using a small stainless steel bone screw. The use of dental cement is also helpful. Attachment of the connector to this location often causes considerably less discomfort to the animal compared to a connector that is sutured to the animal's skin over its back or on its leg. Externalized connectors mounted at the latter locations are a constant nuisance to many animals and often end up chewed loose, with the lead wires destroyed.

## 3.4 *In situ* calibration of buckle transducer

After completing the *in vivo* experimental recordings, the animal must be deeply anaesthetized to carry out an acute non-survival *in situ* calibration of the transducer on the tendon. After exposure of the muscle and tendon, the proximal bony attachment of the muscle must be rigidly clamped and an isometric force transducer attached to the distal insertion of the tendon. It is easiest to cut free the piece of bone to which the tendon attaches, tying the bone and tendon to the isometric force transducer. Because of its low compliance, thick silk suture is good to use for this purpose. When assessment of the muscle's ability to generate torque as a function of joint angle is desired, a considerably more complicated arrangement must be adopted to measure force at the 'output lever' side of the joint (for example, the foot, when measurements of force and torque are being made at the ankle). The joint's axis of rotation must be held fixed while joint moments are measured over a range of joint angles. The buckle transducer can be calibrated by either stimulation of the muscle's nerve using a pulse stimulator (for example, Grass Instruments, Inc.) to develop a maximal isometric tetanus (*Figure 8*) or by directly pulling on the tendon's bony insertion via the isometric force transducer (a Kistler model 9203 piezoelectric transducer was used in the experiment discussed below). In most cases the animal must be sacrificed following *in situ* calibration of the transducer. At this time, measurements of muscle and tendon mass, cross-sectional area, and muscle fibre length can be made. These morphological measurements are needed for calculations of muscle stress, tendon stress and strain, and tendon elastic strain energy storage (see Chapter 3).

## 3.5 Example recordings

Representative recordings from the combined Achilles tendon of kangaroo rats (heteromyid rodents that hop bipedally) are shown in *Figure 9* during hopping and jumping (20). Below each record of muscle–tendon force is the

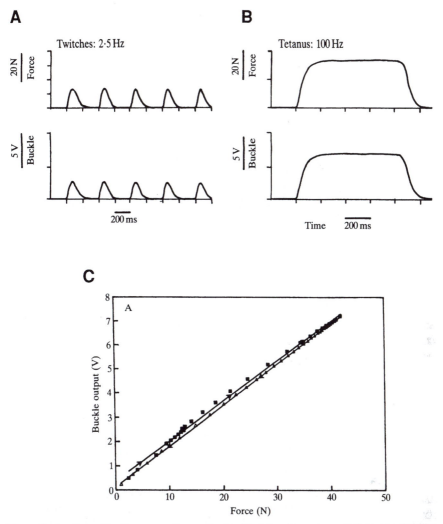

**Figure 8.** *In situ* calibration recordings of a buckle transducer mounted on the Achilles tendon of a kangaroo rat, showing the transducer's voltage output together with that of force measured by an isometric force transducer during a series of (**A**) twitches and (**B**) a tetanus. **C**: A linear dynamic calibration of buckle voltage output versus force is obtained, showing little hysteresis between the rise and decline in force. (Adapted from ref. 20.)

ground reaction force exerted by the animal's limb when it landed on the force platform. In these experiments, the buckle transducers were surgically attached and subsequently recorded from in the space of one day following the animal's recovery from surgery. Measurements of muscle force in this species show that the animal's ankle extensor muscles (gastrocnemius and plantaris) exert about one-third of their peak isometric force when the animals

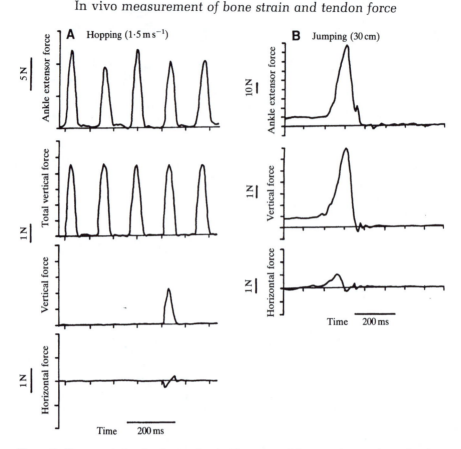

**Figure 9.** Representative *in vivo* tendon (ankle extensor) force and ground reaction force recordings from a kangaroo rat during (**A**) hopping (1.5 m/sec) and (**B**) a jump. (Adapted from ref. 20.)

hop at their preferred speed (1.5 msec$^{-1}$), but are capable of exerting up to 1.75 times peak isometric force during maximal vertical jumps (50 cm high; equal to 10 times their hip height). These very high forces indicate that the muscles were initially stretched for a very brief time, allowing them to develop forces greater than isometric, before shortening to generate the large power output required by the animal to jump to these impressive heights. These high forces also indicate that the nervous system fully activates these muscles to achieve maximal jump heights, a behaviour that the animals perform to avoid prey capture in the wild.

## 4. Summary

*In vivo* muscle–tendon force and bone strain recordings can provide high quality, detailed data for a variety of physical activities, such as animal locomotion. However, the potential for injury because of the invasive nature of these procedures must be weighed against non-invasive (though potentially less accurate or less detailed) approaches, such as those discussed in Chapters 3 and 7, before selecting which approach is best. In the end, if an animal's locomotor performance is critically impaired, no matter how detailed and precise, the data obtained have little biological value.

# References

1. Biewener, A. A. and Taylor, C. R. (1986). *J. exp. Biol.,* **123,** 383.
2. Biewener, A. A., Thomason, J. J., Goodship, A. E., and Lanyon, L. E. (1983). *J. Biomech.,* **16,** 565.
3. Rubin, C. T. and Lanyon, L. E. (1982). *J. exp. Biol.,* **101,** 187.
4. Hylander, W. L. (1979). *J. Morphol.,* **159,** 253.
5. Biewener, A. A. and Blickhan, R. (1988). *J. exp. Biol.,* **140,** 243.
6. Loeb, G. E., Hoffer, J. A., and Pratt, C. A. (1985). *J. Neurophysiol.,* **54,** 549.
7. Walmsley, B., Hodgson, J. A., and Burke, R. E. (1987). *J. Neurophysiol.,* **41,** 1203.
8. Ker, R. F., Alexander, R. M., and Bennett, M. B. (1988). *J. Zool. Lond.,* **216,** 309.
9. Rack, P. M. H. (1985). In *Feedback and motor control in invertebrates and vertebrates* (ed. W. J. P. Barnes and M. H. Gladden), pp. 217–229, Croom Helm, London.
10. Biewener, A. A., Swartz, S. M., and Bertram, J. A. E. (1986). *Calc. Tiss. Int.,* **39,** 390.
11. Goodship, A. E., Lanyon, L. E., and MacFie, H. (1979). *J. Bone Jt. Surg.,* **61A,** 539.
12. Lanyon, L. E., Goodship, A. E., Pye, C. J., and MacFie, H. (1982). *J. Biomech.,* **12,** 41.
13. Rubin, C. T. and Lanyon, L. E. (1985). *Calc. Tiss. Int.,* **37,** 411.
14. Lanyon, L. E. (1976). *Acta Orth. Belg.* **42** (Suppl. 1), 98.
15. Dally, J. W. and Riley, W. F. (1978). *Experimental stress analysis* (2nd edn). McGraw-Hill, New York.
16. Carter, D. R., Harris, W. H., Vasu, R., and Caler, W. E. (1981). In *Mechanical properties of bone* (ed. S. C. Cowin). ASME Publication 45.
17. Carter, D. R. (1978). *J. Biomech.,* **11,** 199.
18. Currey, J. D. (1984). *The mechanical adaptations of bones.* Princeton University Press, Princeton, NJ.
19. Miller, J. M. and Robins, D. (1990). *Ophthalmology & visual science,* **31**(4), 289.
20. Biewener, A. A., Blickham, R., Perry, A. K., Heglund, N. C., and Taylor, C. R. (1988). *J. exp. Biol.,* **137,** 191.

# 7

# Finite element analysis in biomechanics

GARY S. BEAUPRÉ and DENNIS R. CARTER

## 1. Introduction

Most bioengineers would agree that finite element analysis (FEA) has had a profound impact in the field of biomechanics. Finite element analysis provides analysts with a technique to quantitatively analyse complex biological systems for which other solution strategies are either impractical or completely intractable. For example, in the field of orthopaedic biomechanics, FEA has increased our understanding of successful design strategies for total joint replacements to such an extent that the majority of research laboratories, as well as most manufacturers of orthopaedic implants, now consider FEA to be an indispensable analysis and design tool. In addition, FEA is becoming established as a powerful approach to implement and test new theories for the role of mechanical stresses in the development, maintenance, and repair of many tissue and organ systems.

The purpose of this chapter is to introduce some of the basic components of the finite element method as it is applied to structural mechanics, with emphasis on skeletal biomechanics. Although our focus is directed toward finite element analysis of the skeletal system, much of the presentation applies equally as well to other areas of biomechanics such as modelling blood flow, soft tissue mechanics, or heat transfer in biological tissues.

We begin by providing a brief background to the finite element method, and illustrate its use to solve a simple one-dimensional (1-D) example problem. We then examine the basic steps which make up a finite element analysis. Finally, we show some applications of finite element analysis by reviewing example studies of skeletal development and adaptation.

## 2. Background

The finite element method (FEM) is a numerical tool to obtain approximate solutions to complex problems. Most of these problems can be represented in mathematical terms by one or more partial differential equations. While some partial differential equations are amenable to classical analytical solution techniques, most are not. The FEM is a technique to obtain numerical solutions to this class of difficult problems.

In this chapter we focus on the use of the finite element technique to solve problems of solid biomechanics. This chapter is not intended to provide a comprehensive background of this field or a detailed description of the method. Rather, its purpose is to describe in simple terms the essence of the finite element technique as it is used to perform the static analysis of various skeletal structures. For those readers desiring a more complete description of the principles of structural and continuum mechanics and finite element analysis, a classic text in the field is *The Finite Element Method* by Zien-kiewicz (1). A brief review of finite element analysis in orthopaedic bio-mechanics is given by Huiskes and Chao (2).

## 2.1 Concepts and terminology

In the field of solid mechanics one is often interested in determining the displacements (deformations), strains (displacement gradients), and stresses (force intensities) within a structure. In this instance the relevant equations are the equations of static equilibrium from the theory of elasticity. The equilibrium equations are usually written in terms of the stresses or in terms of the displacements. In either case, the resulting equations consist of a set of partial differential equations which must be satisfied at all points within the region of interest. In general, it is not possible to solve this system of equations explicitly except for the simplest of geometric shapes and boundary conditions. In the case of static equilibrium, the finite element technique can be thought of as replacing this complex (and often intractable) system of partial differential equations with a simpler system of algebraic equations. This simpler system of algebraic equations is obtained by dividing the struc-ture (often referred to as the 'problem domain') into a number of subregions or elements which are interconnected at discrete points or nodes. The process of dividing the structure of interest into nodes and elements to create a finite element mesh is called discretization. The solution of the finite element equations no longer satisfies the equilibrium equations at each and every point within the domain of the original problem. Rather, the governing equations for the displacements are satisfied only approximately at specific locations referred to as nodal points. The finite element technique thus involves the replacement of a complex system of partial differential equations having an infinite number of degrees of freedom with a discrete system of algebraic equations having a finite (although often large) number of degrees of freedom.

The system of algebraic equations which is solved during a finite element analysis is given as:

$$\{F\} = [K]\{u\} \tag{1}$$

where $\{F\}$ is the vector of nodal forces, $[K]$ is the element stiffness matrix, and $\{u\}$ is the vector of nodal displacements. To establish the element stiffness matrix, one must know the mechanical properties of the material being modelled. In the case of a simple linear isotropic material, these properties are the Young's modulus and the Poisson's ratio (see Chapter 1 for an introduction to these and other material properties).

The similarity between equation [1] and a force–spring system is readily apparent. Given specific boundary conditions in the form of prescribed nodal forces and displacements, the set of equations [1] can be solved for the remaining unknown nodal displacements. In addition to the displacements, the stresses and strains within a material are also frequently desired. The stresses and strains are easily determined once the nodal displacements are known using the appropriate relations from the theory of elasticity. These quantities represent the solution to the discrete finite element equations, as well as an approximate numerical solution to the original partial differential equations.

## 2.2 Illustrative example of a compact bone specimen subjected to uniaxial loading

This example illustrates the use of the finite element technique to determine the deformation and state of stress in a specimen of compact bone loaded in uniaxial tension. The geometry of the bone specimen is assumed to be in the form of a tapered beam (*Figure 1a*). *Figure 1b* shows a finite element representation of the cortical bone specimen. The finite element model consists of six elements and seven nodes. Each element is described in terms of a length, a material property (Young's modulus) and a cross-sectional area.

For this 1-D example, it is also possible to solve analytically for the displacements and stresses using the equations from classical beam theory (see Chapter 1). The 'exact solution' for this problem consistent with the 1-D approximation is indicated by the solid curves in *Figures 1c* and *1d*. The finite element solution to this problem is also shown in *Figures 1c* and *1d*. The seven nodal displacements calculated using the finite element technique are indicated by the solid circles in *Figure 1c*. The stresses within each element are indicated by the dotted lines in *Figure 1d*. Several features of the finite element solution should be noted:

- there is excellent agreement between the displacements calculated analytically and the nodal displacements provided by the finite element solution;
- the stresses calculated from the finite element model are constant within each element;
- the finite element stresses are discontinuous between adjacent elements.

The lack of stress continuity across element boundaries is a feature common to finite element solutions. The magnitude of the stress discontinuities

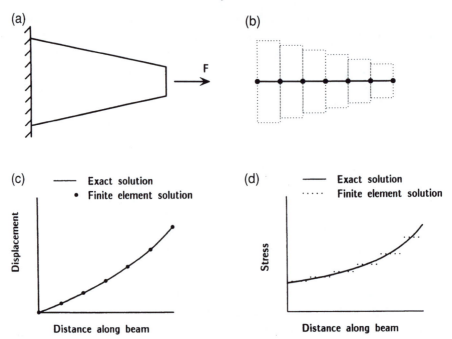

**Figure 1.** (a) A specimen of cortical bone in the form of a tapered beam. The bone specimen is rigidly fixed at the left end and subjected to an axial force, *F*, at the right end. (b) A finite element representation of the bone specimen consisting of 7 nodal points and 6 one-dimensional elements. (c) Graph showing the variation in displacement along the length of the specimen. The finite element results are shown as solid circles. (d) Graph showing the variation in stress along the length of the specimen.

serves as an indication of the accuracy of the solution. It is easy to imagine that as the number of nodes and elements is increased the stress discontinuities will become smaller and smaller. Whereas the classical analytical solution of this beam problem consists of a continuous description of the displacements and stresses at each and every point within the beam, the discrete finite element solution consists only of the displacements at the seven nodes and various derived quantities such as the stresses within each of the six elements.

## 3. Basic steps in a finite element analysis

The application of the FEM to any problem in structural analysis typically involves three basic phases:

- model creation
- solution
- results validation and interpretation

The finite element process is best understood by examining each of these phases separately and in detail.

## 3.1 Model creation

The goal of the model creation phase is the mathematical description of the finite element model in term of nodes and elements, material properties, boundary and interface conditions, and applied loads. The model creation phase is often referred to as the preprocessing phase, since it takes place prior to the processing or solution phase. The model creation phase can often be the most labour-intensive phase of a FEA. Complicated 3-D models can sometimes take weeks or even months to create. In recent years, special-purpose computer programs called finite element preprocessors have greatly simplified the process of generating finite element models. Although many finite element programs include a preprocessor with mesh generation capabilities, they tend to be much less sophisticated than some of the stand-alone interactive programs that were developed expressly for the purpose of creating finite element models. One such program is PATRAN II. By using different translator programs, a preprocessor program can generate finite element input data in the proper format for a variety of FEA programs.

### 3.1.1 Elements and nodes

The analyst is faced with many choices during the creation of a finite element model. One must decide first whether to describe the system using a 1-D, 2-D, or 3-D representation. The initial choice will dictate the types of elements which can be used to discretize the problem domain. In the previous example of the tapered beam, 1-D elements were used. The physics of the problem being modelled will determine which type of element is most appropriate. In biomechanics, the representation of structures with 1-D elements is rarely adequate. However, a 2-D description may be perfectly acceptable. A plane stress or plane strain analysis is particularly useful when the important deformation of a structure occurs primarily in one plane. Another type of 2-D approach is based upon the assumption of axisymmetry. Axisymmetry implies that the geometry has rotational symmetry about one axis. Even though these conditions in skeletal biomechanics are rarely met in a strict sense, two-dimensional models are still widely used.

The process of defining a structure in terms of nodes and elements is referred to as 'meshing'. All two-dimensional regions are 'meshed' using either triangular or quadrilateral elements. Some typical 2-D elements are shown in *Figure 2a–d*. These elements can have two or more nodes along each side of the elements. Elements having two nodes per side are referred to as linear elements, because linear polynomials are used to define the element geometry. Elements having three nodes per side are referred to as quadratic elements. Quadratic elements are, in general, more accurate than linear elements, but are computationally more expensive (that is, they require more

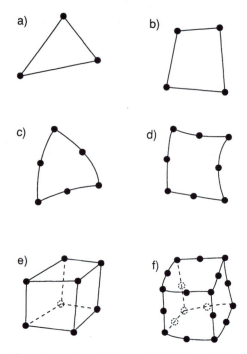

**Figure 2.** Examples of common elements types. (a, b) Linear triangular and quadrilateral elements for meshing two-dimensional regions. (c, d) Higher order, quadratic elements for meshing two-dimensional regions. (e, f) Linear and quadratic three-dimensional 'brick' elements.

computer time per element). Elements having more than three nodes per side are almost never used. When meshing two-dimensional regions, quadrilateral elements traditionally have been the preferred element type. The exclusive use of linear (three node) triangular elements is generally discouraged, since finite element developers have shown that finite element models constructed using these particular elements are not as accurate as models that are based on other element types. In recent years, automatic mesh generators have relied more heavily on the use of quadratic six-node triangular elements because of the ease with which they can be used to represent complicated shapes.

Some typical 3-D elements are shown in *Figure 2e, f*. Once again, these include linear and quadratic elements. The most common 3-D elements are the linear eight-node brick and the quadratic twenty-node brick. Three-dimensional wedge and tetrahedral elements are also available. As with the use of linear triangular elements, the use of linear wedge and tetrahedral elements is generally discouraged.

The finite element method has been used to study problems in biomechanics for nearly twenty years. The vast majority of analyses during the first fifteen of those years used linear 2-D models. As faster computers and larger capacity disk drives have decreased in cost, more and more studies now use three-dimensional and/or non-linear models. Nevertheless, two-dimensional models still play a key role in the analysis of many problems. In particular, parametric studies that analyse a model repeatedly, using different loads or material properties, can be significantly less costly to perform using a 2-D model. The results from a 2-D model are also usually much simpler to interpret. The sheer volume of data associated with a 3-D model can make both visualization and interpretation of the results a difficult and time-consuming process.

The process of choosing between a 2-D and a 3-D modelling approach must be made on a case by case basis. The analyst must weigh the benefits of a 3-D analysis against the higher costs of making, analysing, validating, and interpreting a 3-D model. In general, the benefits of a 3-D approach should usually be assessed in terms of the limitations of the alternative 2-D approach. Conventional plane stress and plane strain analyses do not include the effects of out-of-plane stiffness. Pseudo 3-D approaches (3, 4), which lie somewhere between 2-D and 3-D approaches, have been used with reasonable success. The two basic questions that the analyst must ask when deciding between a 2-D and a 3-D modelling approach are: first, what mechanical characteristic does the third dimension provide? and second, is the inclusion of that mechanical characteristic necessary for the problem at hand? A good background in solid mechanics and experience with finite element analysis can help the analyst answer these questions. Nevertheless, there are times when the deficiencies in a 2-D model are apparent only after a 3-D model has been analysed.

### 3.1.2 Material properties

The specification of material properties is one aspect of FEA which can be particularly difficult when modelling biological systems. Many biological tissues have material properties which vary with location (inhomogeneous), direction (anisotropic), loading rate (poroelastic, viscoelastic), and load (non-linear).

Most FEAs which have appeared in the biomechanics literature have modelled bone as a linear elastic material. This is a reasonable assumption under non-impact loading conditions. The material characterization of bone is none the less complicated by other considerations. First of all, the properties depend strongly on bone density, which can vary widely from one location to another (inhomogeneity). Carter and Hayes (5) developed an empirical relationship between elastic modulus and bone density of the form:

$$E \propto \rho^3. \qquad (2)$$

This relationship is obviously a simplification of the actual situation, since density alone does not completely determine the elastic modulus of bone. In both cancellous and cortical bone, the microstructural organization of the bone also has an important influence on bone stiffness and strength. The microstructural arrangement often leads to different stiffness and strength characteristics in different directions (anisotropy). For example, the cortical bone in the shaft of the human femur is nearly 50% stiffer in the axial direction than in the transverse or circumferential directions (6). Cancellous bone found in the ends of long bones and in vertebral bodies, can also be highly anisotropic. The anisotrophy is due to the orientation of the individual trabeculae which make up all cancellous bone. Although isotropy (which involves two independent material constants; for example, Young's modulus, $E$, and Poisson's ratio, $v$) is the most common material model, more realistic (and complicated) material models are also used.

The two most common anisotropic material models for bone are transverse isotropy (five independent material constants) and orthotropy (nine independent material constants). The term 'transverse isotropy' implies that the material behaves isotropically in the 'transverse' plane. Haversian cortical bone from the diaphysis of a long bone (which has a characteristic longitudinal direction defined by the axis of the osteons) is well-represented as a transversely isotropic material. In this case the transverse plane is perpendicular to the long axis of the bone and in this plane the behaviour is assumed to be isotropic; that is, the radial and circumferential moduli are assumed to be equal. For a transversely isotropic material the five independent material constants are usually represented by two Young's moduli (one longitudinal—that is, parallel to the osteonal axis—and one transverse), two Poisson's ratios, and one shear modulus. An orthotropic material with nine independent constants can be described by three Young's moduli, three Poisson's ratios, and three shear moduli. Cancellous bone is sometimes represented as an orthotropic material in order to describe better its directional material behaviour. The most general description of a material requires the specification of 21 independent constants. However, this degree of complexity is almost never necessary. Any of these representations is possible with modern finite element programs.

### 3.1.3 Loads, boundary, and interface conditions

The final task in the model creation phase is the specification of loads, boundary conditions, and interface conditions. When modelling parts of the skeletal system, the loads of which we are speaking typically are the joint forces, musculotendon forces, and ligament forces that exist during gait. The determination of the time history of these loads has long been a goal of orthopaedic research. Much of this research has focused on the loads which act on the human femur and, in particular, on the proximal femur and femoral head. Early attempts at determining the loads at the hip were based on simple kinematic models. More recently, dynamic muscle-optimization models and load-sensing implants have been used to determine more completely and accurately physiologic load magnitudes and directions. The use of *in vivo* strain-gauges has also provided much information about the state of strain in limb bones of a number of animals studied (see Chapter 6).

As our knowledge of physiological loads increases, the skeletal models that are created tend to become more complicated. For example, recent studies based on telemeterized recordings of *in vivo* loading from hip prostheses in humans (7, 8) indicate that the forces acting on the proximal femur do not always lie in the mediolateral plane. Substantial out-of-plane forces causing both bending and torsion also exist, particularly during activities such as rising from a chair and stair-climbing. The existence of these forces indicate an important limitation inherent in any two-dimensional model of the human femur.

Boundary conditions in finite element models consist of displacement constraints which are required in any static FEA to prevent rigid body motion of the model. These constraints can be representative of symmetry or anti-symmetry conditions or simply as prescribed nodal displacements (usually taken to be zero). The displacements are typically applied at sections of the model far removed from the primary areas of interest. This practice is prudent, since prescribed displacements can create stress artefacts in the vicinity of the constraints. Symmetry and anti-symmetry conditions should be used whenever appropriate (for example, sagittal symmetry in the vertebrae) since they result in a great saving in computational effort.

For structures constructed of differing materials or material organization, the conditions which exist at the interfaces between these regions in a model can greatly affect the predicted response of the system. For example, the modelling of a prosthetic joint might include regions corresponding to bone, cartilage, metal, cement, and polyethylene. The simulation of fully-bonded interfaces having displacement continuity at the interface is the easiest to model since these interface conditions do not violate the assumptions of linearity. On the other hand, gap and frictional interfaces which can separate under tensile loading require a non-linear, iterative solution technique. Most, but not all modern finite element programs have the capability of modelling either linear or non-linear interface conditions.

## 3.2 Solution

The solution phase consists of executing a finite element computer program using the previously generated model as the input data. Because most FEAs are computer-intensive (requiring extensive computer time), program execution usually takes place as a batch or background process on the computer, rather than as an interactive process. Generally, the entire solution phase occurs without any interactive action on the part of the analyst. The amount of computer time required to complete the analysis is a function of the number of nodes in the model and the number of degrees-of-freedom per node. In the course of achieving a solution, most finite element programs perform a series of checks of the model to ensure that the model is correctly defined. The finite element program may check for proper data format, element definitions, adequacy of the boundary conditions, the existence and appropriateness of material properties, etc. The solution phase is complete when the nodal displacements and any derived quantities (stresses, strains) have been calculated and stored in digital form on the computer.

## 3.3 Validation of results

A crucial part of any FEA is the validation of the results. As pointed out by Huiskes and Chao (2), the validation of any FEA involves an assessment of two distinct issues: model validity, and model accuracy. Model validity is concerned with the degree to which the finite element model represents the real system being modelled. In other words, does the finite element model adequately represent the geometry, loads, material properties, boundary, and interface conditions of the real structure. If any key feature of the physical system is not included in the finite element description, then model validity is compromised and some aspect or response of the model will be lacking. Neglecting an important load, specifying inappropriate boundary conditions, or choosing improper material properties are just a few of the many ways in which the model may be deficient or incorrect.

Once a valid model has been created, its accuracy can be assessed using a convergence test. A convergence test involves refinement of the finite element mesh (increasing the number of nodes and elements in critical areas of the model) and subsequent re-analysis of the model to determine how increasing the mesh density (decreasing the size of the elements) affects the results or predictions of the model. The goal of this refinement process is to increase the accuracy of the finite element solution with each subsequent refinement. Ideally this refinement process is carried out repeatedly until some critical solution variable (for example, peak stress, strain, or strain energy) does not change significantly from one refinement to the next. In the past, convergence studies have been performed only infrequently. The extensive time and effort required of the analyst to refine the finite element mesh and the increase in computer time and storage space are cited as the primary reasons why analysts do not perform convergence studies. Fortunately, recent advances in computer hardware and finite element programs may change this. Some finite element programs (9) can now perform a convergence study automatically. With these programs, the user is responsible for developing an initial finite element model. After the finite element program analyses the initial model, the program refines the mesh and then analyses the refined model. This process is repeated until some predetermined convergence criteria are met. The automated nature of this process makes it much more likely that analysts will perform convergence studies in the future.

Whenever possible, the validation process should also include a comparison with existing data or with other evaluation techniques. This might include the following:

- strain gauge data from *in vivo* and *in vitro* tests;
- data from *in vivo* load-sensing implants;
- comparison with other analytical solution techniques such as elasticity or beam theory; or
- comparison with bone and tissue morphology; for example, in regions where high magnitude stresses are predicted, is denser bone observed?

Although some of these items might seem obvious, it is surprising how often these validation checks are not performed.

### 3.3.1 Interpretation of results

Because of the vast quantity of data which is generated in a FEA, the interpretation of results can be a difficult and time-consuming task. The first step in the interpretation process is the visualization of results. The use of an interactive, graphics-based post-processing program is the most efficient way to visualize results. Many finite element program developers include some post-processing capabilities, either integrated within the finite element program or available as a separate companion program. Independent post-processing programs (like PATRAN II) which have interfaces or translators to a variety of finite element programs can also be used.

For problems in hard-tissue biomechanics, the displacements, stresses and strains within the structure are the variables that are most frequently of interest. The graphical visualization of displacements is usually done by displaying the finite element model or mesh in the deformed configuration. Since the magnitudes of the displacements are typically quite small, a magnification factor is often used to exaggerate the deformations.

Contour plots, colour fringe plots and vector plots are the most common ways to display stress and strain information. Contour and fringe plots are used to visualize stress or strain on a surface or plane of the model. Contour plots can be made of any of the six individual normal and shear stress components. Scalar quantities such as von Mises' stress or strain energy density are also displayed in this way. Vector plots are used to display principal stress or strain magnitudes and directions.

There are several pitfalls that can occur in the interpretation phase of a FEA. The most common pitfall in orthopaedic biomechanics is overemphasizing the precise stress or strain values in a model. The analysts must keep in mind that all finite element models of bones are idealized to some extent. Some 3-D models that have appeared in the literature have used meshes that are very coarse. In addition, three-dimensional models are almost never validated using a convergence study. The analyst should also be aware of the

159

sensitivity to model parameters such as loading conditions, boundary conditions, and material properties. In most cases, finite element results should be used primarily to indicate trends or general mechanical behaviour. Results obtained with different model parameters should be compared with one another and not to some assumed absolute standard.

Finally, the results of a finite element analysis should be interpreted in terms of meaningful measures such as percentage of local failure stress or strain. By interpreting results in this way, the analyst can answer some key questions, such as: do the results imply that the bone will fracture or the prosthesis will loosen during walking; or, do the results imply something that is otherwise unreasonable or impossible?

# 4. Applications

The ways in which the finite element technique has been applied in biomechanics has changed significantly since the method was first introduced to biomechanics nearly 20 years ago. To appreciate the nature of this change one must understand how the technique was originally used and, for the most part, how it still is being used in other engineering disciplines. In other fields in which the FEM is used, the main goal of an analysis might be the determination of the displacements, stresses and strains within some structure being modelled. For example, in civil or aerospace engineering it might be sufficient to know the stresses within a material or the displacement at some key location within a structural component. The calculated stresses or displacements might then be compared with a set of predetermined failure criteria. At this point the analysis might be considered complete. In biomechanics, a similar approach is sometimes taken; the goal being the determination of the stresses and strains within a bone or within some orthopaedic device, such as an artificial joint or metal implant. This type of analysis can be thought of as mainly a structural analysis applied to problems in orthopaedics or biology more generally. While these types of studies are important, the finite element technique can also be used with broader goals in mind. In recent years, the FEM has been used to gain insight into various biological processes. This type of analysis is potentially very rewarding since an understanding of the mechanisms which underlie certain biological processes may lead to a complete re-evaluation of existing treatment modalities, orthopaedic devices, and surgical techniques.

The remainder of this chapter is devoted to a series of computer simulations which focus on the role of mechanical stress in skeletal development, growth, and adaptation. In each of the following simulations we use FEA as a tool to quantify the mechanical state of stress within the skeletal elements in question. The FEA represents only one aspect or phase of each of the simulations that we present. The end goal of each in these simulations is not the determination of the state of stress within the bone, but rather to achieve a better understanding of how the stress state might influence or modulate the biological response of the system.

The first simulation addresses the influence of mechanical stress on the morphogenesis of the human femur during fetal and postnatal development. The second simulation focuses on the subsequent bony adaptation of the proximal femur. Finally, the third simulation concerns the influence that different load histories have on the morphology of the femoral diaphysis. Each of these simulations was conducted on a VAX/VMS 11/780 minicomputer using the finite element program ABAQUS and the pre- and post-processor PATRAN II.

## 4.1 Morphogenesis of the fetal and postnatal femur

During embryonic development, the position and shape of future bones are established initially by cartilaginous condensations called anlagen. The subsequent growth and chondro-osseous development of the anlagen are influenced by multiple extrinsic factors in the immediate vicinity of the anlagen. These factors include the influence of adjacent tissues, vascularity, nutrient and hormone concentrations, and mechanical loads.

*In utero*, mechanical loading of the anlagen occurs through muscular activity of the developing fetus and through external loading imparted by movements of the mother. The appearance of the primary ossification site occurs at roughly the same time as the first sign of contractile activity of the developing muscles. Ossification of the femoral anlage begins at the centre of the femoral diaphysis and progresses away from the centre toward the bone ends. Two to four months after birth, a secondary ossification nucleus appears in the centre of the proximal femoral epiphysis. Ossification then proceeds around the secondary ossific nucleus until the only remaining cartilaginous areas are the articular cartilage and the growth plate between the primary and secondary ossification fronts.

Our research group (10, 11) has developed a hypothesis that the natural temporal sequence of events in the cartilage anlage is proliferation, maturation, hypertrophy, and ossification. Additionally, we hypothesize that intermittently imposed shear stresses will speed up this sequence of events, while intermittently imposed compressive hydrostatic stresses will retard or arrest this sequence. Fundamental to this hypothesis is the notion that cartilage is nearly incompressible. As a result, hydrostatic pressure will create negligible deformations and therefore negligible stored strain energy in the cartilage. The main contribution to the stored strain energy comes from the shear deformation. Regions of high strain energy are therefore equivalent to regions of high shear stress and, according to our hypothesis, are the regions of the anlage that will ossify first. By performing a finite element analysis and examining the distribution of strain energy, we can predict which regions of the cartilage anlage will be the first to ossify.

161

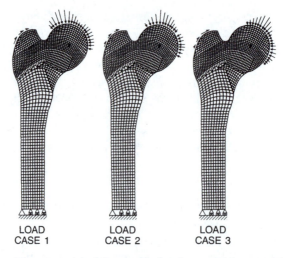

LOAD CASE 1    LOAD CASE 2    LOAD CASE 3

**Figure 3.** Finite element model of the developing femoral anlage and the three loading conditions used to simulate the loading history. (From Carter *et al.*, ref. 10.)

To test this hypothesis we created a plane strain two-dimensional finite element model of the fetal proximal femur. Although the loading conditions in the embryo, fetus, and neonate are unknown, we assumed that the loading history imposed on the proximal femur could be represented by three loading conditions which were intermittently applied for an equal number of loading cycles per day (*Figure 3*). The model represents the proximal femur from 48 days after conception to approximately 18 months after birth. The model consists of approximately 1400 quadrilateral and triangular elements. The shape of the model was based upon the proximal femur at four months after birth as determined by Trueta (12). The model was digitized from a photograph and meshed and analysed using the GIFTS finite element program (see Appendix).

The material properties used in the model were based on empirical values reported in the literature. Cartilage was modelled as a homogeneous, linearly isotropic material with an elastic modulus of 6 megapascals (MPa) and a Poisson's ratio of 0.47. The initial analysis was conducted with all elements having the material properties of cartilage. Later stages of development were simulated by changing the properties of the ossified regions to those of bone. All ossified regions are assumed to have a modulus of 5000 MPa and a Poisson's ratio of 0.35.

Several points should be noted about this model. First of all, it is a two-dimensional model and therefore it is not possible to represent out-of-plane or torsional loads. However, for the proximal femur we felt that these additional loading conditions would not change the basic patterns of ossification which we anticipated with a two-dimensional representation. Secondly, the finite element mesh was relatively refined (that is, the elements were relatively small) so that a convergence study was not considered mandatory. The finite element novice should be aware that considerable experience with FEA is necessary before making these types of assumptions. Even with adequate finite element experience, it is not always possible to predict accurately the effects of some modelling assumptions.

The distributions of strain energy density are shown in *Figure 4a–d* at four different stages of ossification of the femoral anlage. Each of these four different stages was represented using a separate finite element model. The results were obtained by adding the strain energy density values calculated under the three different loading conditions. For the 48-day model (*Figure 4a*), the highest strain energies occur on the periosteal surface at the midshaft. This location corresponds with the initial formation of the tubular diaphysis seen *in vivo*. As ossification progresses toward the bone's end, the highest magnitude of strain energy is immediately ahead of the ossification front (caused by the difference in stiffness between cartilage and advancing bone) and still highest at the periosteal surface (*Figure 4b, c*). When the ossification front approaches the epiphyseal region a transition occurs. The region of highest strain energy suddenly shifts from the advancing mineralization front to the centre of the chondroepiphysis (*Figure 4d*). This corresponds to the appearance of the secondary ossification centre. These progressing regions of high strain energy are completely consistent with the ossification sequence observed *in vivo*.

In each stage of ossification represented in *Figure 4a–d*, compressive stresses exist near the joint surface in the femoral head. This stress state is indicative of hydrostatic pressure and relatively low strain energy. Thus, we would

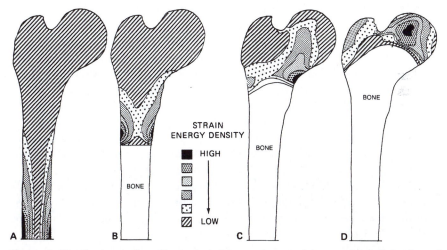

**Figure 4.** Distributions of cartilage strain energy imposed by mechanical loading at various stages of ossification. (From Carter *et al.*, ref. 10.)

predict that this region would remain as cartilage and undergo degeneration and ossification only later in life, corresponding to degenerative joint disease in older adults.

## 4.2 Bone adaptation in the proximal femur

This second application utilizes another theory developed by our research group which enables us to simulate the time-course of bone adaptation in skeletal elements. A comprehensive description of the theoretical foundation for the algorithm is available for the interested reader (13). A brief overview of the approach is presented here.

The remodelling algorithm is based upon the notion that bone tissue requires a certain level of mechanical stimulation (stress stimulus) to maintain itself. If bone experiences excess stimulation, additional bone tissue will be deposited. If bone experiences insufficient stimulation, it will resorb. We assume that the appropriate level of stimulation necessary for bone maintenance is determined by genetic and systemic factors and can be modulated by interactions with adjacent tissues. The appropriate level of stimulation is called the attractor state stress stimulus. The term 'attractor state' is borrowed from the field of non-linear dynamics and is meant to represent a target state toward which some types of dynamic, time-varying systems (like the remodelling skeletal system) are attracted. The attractor state stress stimulus for bone tissue has been estimated from various *in vivo* studies reported in the literature (14).

The level of mechanical stimulation actually experienced by bone tissue is a function of its daily loading history. To quantify the loading history we define a daily stress stimulus as:

$$\psi_b = \left[ \sum_{\text{day}} n_i \bar{\sigma}_{b_i}^m \right]^{1/m},\tag{3}$$

where $n_i$ is the number of cycles of load type $i$, $\bar{\sigma}_{b_i}$ is a measure of the effective stress at the tissue level, and the exponent $m$ is an empirical constant.

Each load type, $i$, in equation (3) represents a different physical activity; for example, walking, rising from a chair, stair climbing, running, etc. The number of cycles per day, $n_i$, for each load type is estimated from the literature.

The daily stress stimulus varies spatially within the bone depending upon the local state of stress and strain. Finite element modelling is used to quantify the stresses and strains. The rate of bone remodelling is determined by the difference between the level of stimulus a region of bone actually experiences and the appropriate level required for bone maintenance. When this difference is large, the rate of remodelling is large. When this difference is small, the rate of remodelling is small or zero.

This remodelling algorithm is shown schematically in block diagram form in *Figure 5*. In this representation, the load history acts on the bone to produce a

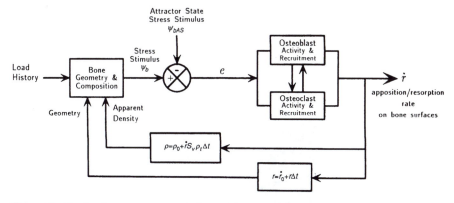

**Figure 5.** Block diagram representation of bone remodelling. (From Beaupré *et al.*, ref. 13.)

stress stimulus. This stress stimulus is compared with the constant attractor state stress stimulus to produce an error signal. The error leads to changes in the recruitment rate of osteoblasts and osteoclasts that in turn alters the rate of bone modelling and remodelling. The rate of surface apposition or resorption is then used to update the external geometry and bone density. The entire system is thus a dynamic system that is kept in balance by multiple feedback control mechanisms.

The two-dimensional finite element model utilized in the present simulation is shown in *Figure 6*. A two-dimensional representation was used in order to permit a parametric examination of loads and remodelling constants. Subsequent preliminary results using a 3-D model (15) indicate that the two-dimensional predictions are not unrealistic when modelling a mid-frontal section of the femur. However, the three-dimensional study did point out the importance of out-of-plane loads on the predicted three-dimensional morphology of the femoral diaphysis. If the 3-D morphology of the shaft is of primary interest, then a 3-D model would be required.

The loading and boundary conditions used with the two-dimensional model are also shown in *Figure 6*. The first loading condition corresponds to the one-legged stance phase of gait. The other two loading cases were selected to represent extreme ranges of motion of abduction and adduction with reduced force levels. In this simulation, the external geometry was not permitted to change; that is to say, growth was not simulated.

All bone elements in the model were initially assigned an elastic modulus, $E = 500\,\text{MPa}$ (corresponding to a bone apparent density of $0.57\,\text{g/cm}^3$) and a Poisson's ratio, $\nu = 0.2$. The model has a layer of cartilage elements at the joint surface with a shear modulus, $G = 2.0\,\text{MPa}$ and a Poisson's ratio, $\nu = 0.47$.

After the stress fields in the initial, homogeneous model are calculated, the stresses in each bone element are expressed in terms of the effective stress.

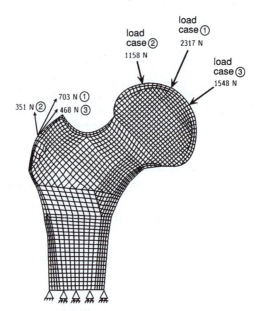

**Figure 6.** Two-dimensional finite element mesh of the proximal femur. Three distributed loading conditions were used. (From Beaupré *et al.*, ref. 14.)

Using the remodelling theory described above, the remodelling error is then calculated. The remodelling error determines the remodelling rates or the rates of osteoclastic resorption and osteoblastic deposition. These in turn are used to update the bone apparent density and material property distributions. The analysis is then repeated using the updated material properties. Each cycle represents one time-increment in the adaptive process of remodelling bone.

Predicted changes in the density distribution after 1, 15, and 30 time-increments are shown in *Figure 7a, b, c*. The sequence of changes show a consolidation of bone in key regions leading to the generation of two cortices and a relatively dense strut of cancellous bone in the femoral head along the primary load path. The area of reduced density, consistent with a similar region in the actual femur referred to as Ward's triangle, is also apparent.

This simulation indicates that many of the morphological features of the proximal femur can be predicted using a stress-based remodelling theory. The algorithm that was used does not require the use of a site-specific value for the attractor state constant; that is, the attractor state had a single value for all locations within the bone. This implies that the loading history alone plays a major role in the determination of the bony architectural features of the proximal femur.

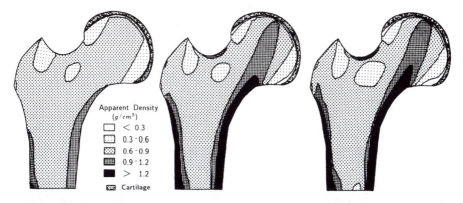

**Figure 7.** Predicted distributions of bone apparent density in the proximal femur after 1, 15, and 30 time-increments. (From Beaupré *et al.*, ref. 14.)

## 4.3 Morphology of the femoral diaphysis

The final simulation illustrates the influence of loading history on the morphology of the femoral diaphysis. The bone adaptation algorithm outlined in the previous application is used in this simulation as well. The goal of the simulation is to examine the predicted morphology of a typical long bone subjected to different assumed loading histories. *Figure 8* shows an idealized model of a long bone diaphysis. As with the previous simulation, we assume that the model initially has a homogeneous material property distribution and that no medullary cavity exists.

Multiple loads, each with a prescribed number of load cycles, were created to simulate various assumed daily loading histories. The loading modes included bending with combined axial compression and torsion (*Figure 9a, b*). These loading modes were used to create three loading histories. Loading history No. 1 consisted of 10 000 cycles per day of bending in an isolated plane with superimposed axial compression. Loading history No. 2 consisted of

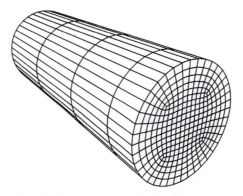

**Figure 8.** Three-dimensional finite element mesh of an idealized long-bone diaphysis.

167

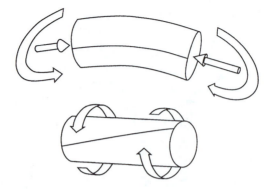

**Figure 9.** Loading conditions for the long-bone diaphysis model. Combined bending and axial compression is shown at the top. Torsion is shown at the bottom.

8000 cycles of bending with superimposed axial compression and 2000 cycles of torsion. Loading history No. 3 consisted of 5000 cycles of bending with superimposed axial compression and 5000 cycles of torsion. The load magnitudes were chosen to result in periosteal axial strains of 500 $\mu\varepsilon$ compression with 300 $\mu\varepsilon$ tension and shear strains of 525 to 935 $\mu\varepsilon$.

The predicted distributions of apparent density for loading history No. 1 are shown in the top row of *Figure 10*. Results are shown for time-increments 0, 2, and 15. After 15 time-increments, the resulting cross-section resembles a non-symmetric I-beam with a thicker flange at the top (compression cortex). For this loading history there is no contiguous cortex that connects the tension and compression cortices, which is clearly an unrealistic solution. The predicted density distribution for loading history No. 2 is shown in the middle row of *Figure 10*. The addition of a torsional load component now leads to a contiguous cortex with a thicker compressional aspect. Finally, the density distributions for loading history No. 3 are shown in the bottom row of *Figure 10*. Once again, a contiguous cortex is predicted which is only slightly thicker on the compression side.

This simulation shows that different loading histories result in different predicted cross-sectional morphologies in a long bone. Axial compression combined with bending in a single plane is not capable of generating a contiguous cortex. By restricting the bending to act in a single plane, the bone achieves a highly optimized I-beam structure that would be inefficient for resisting bending in other planes or torsion. Multiple loading modes and/or multiple planes of bending leads to a structure capable of withstanding different types and directions of loads.

## 5. Summary

The simulations presented in the previous section were chosen to illustrate the use of finite element analysis in biomechanics. In many ways, these

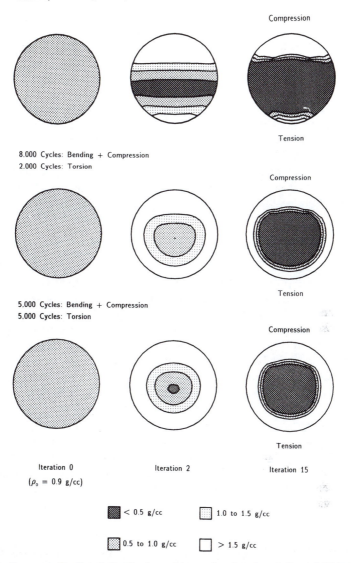

**Figure 10.** *Top row:* Predicted distribution of bone density after 0, 2, and 15 iterations, assuming that the loading history consists of 10 000 cycles per day of axial compression combined with bending in a single plane. *Middle row:* Predicted distribution of bone density, assuming that the loading history consists of 8000 cycles per day of axial compression and bending and 2000 cycles per day of torsion. *Bottom row:* Predicted distribution of bone density, assuming that the loading history consists of 5000 cycles per day of axial compression and bending and 5000 cycles per day of torsion.

applications are relatively simplistic. Nevertheless, they show that FEA can and should be used for more than simply quantifying stresses and strains. This is true in the field of biomechanics in particular and in the field of biology in general. Given the uncertainties with regard to loading and material properties, one should view FEA as a qualitative or at most a semi-quantitative tool. That is not meant to imply that FEA need play a lesser role in biology than in other more traditional engineering fields. It is only meant to suggest that analysts are beginning to appreciate that a simple determination of stresses and strains in a biological structure is generally not of interest in and of itself. There must be some biological or clinical relevance to motivate and guide any FEA. Without this, the process may quickly become an exercise in abstract engineering. A comparison of the use of FEA in biomechanics today with its use twenty years ago, shows that it has become a much more relevant, accessible, and valuable analysis tool.

# 6. Appendix: finite element programs

One of the most exciting developments in recent years is the availability of finite element programs that run on relatively inexpensive personal computers (PCs) (e.g. DOS-based 8086, 80286, 80386, and 80486 systems, and Apple Macintosh II). Access to a large mainframe computer is no longer a prerequisite for performing a finite element analysis. In a 1986 article entitled 'Finite element codes for microcomputers—a review', J. Mackerle (16) listed 41 microcomputer-based finite element programs developed around the world. Some of the programs listed in this earlier article were developed at universities and intended only as educational tools. Other programs listed undoubtedly no longer exist. Still others have evolved into full-featured finite element programs, having modelling capabilities previously possible only with programs running on larger minicomputers and mainframe computers.

The following list of computer programs was compiled primarily from advertisements in various US engineering trade journals. The list is by no means comprehensive. Undoubtedly many more excellent finite element programs exist world-wide. The list is divided into two categories. The first category includes programs that are often referred to as general-purpose or full-featured programs that traditionally have been used on computers larger than PCs. In fact, at the time of this writing, a PC version exists only for one program (ANSYS) of the six programs listed. The second category includes programs that will run on machines as small as personal computers.

Because of the multitude of different finite element programs and different computers, it is impossible to provide minimum computer requirements appropriate for all finite element programs. Having said that, the requirements for the Cosmos/M program are probably not untypical. Minimum requirements for a DOS-based 80386 system include a 40 Mbyte hard disk, 4 Mbytes of RAM, a math co-processor, 1 serial port, 1 parallel port, and an

Category 1: *Finite element programs available on workstations, minicomputers, mainframes, and super-computers*

| *Program name* | *Program developer* |
|---|---|
| ABAQUS | Hibbit, Karlsson & Sorensen, Inc.<br>100 Medway St.<br>Providence, RI 02906-4402<br>(401) 861-0820 |
| ADINA | Adina Engineering<br>71 Elton Ave.<br>Watertown, MA 02172<br>(617) 926-5199 |
| ANSYS | Swanson Analysis Systems, Inc.<br>Johnson Rd., Box 65<br>Houston, PA 15342-0065<br>(412) 746-3304 |
| I-DEAS | SDRC<br>2000 Eastman Drive<br>Milford, OH 45150-2789<br>(513) 576-2400 |
| MARC | Marc Analysis Research Corp.<br>260 Sheriden Ave.<br>Palo Alto, CA 94306<br>(415) 326-7511 |
| MSC/NASTRAN | MacNeal-Schwendler Corp.<br>815 Colorado Blvd.<br>Los Angeles, CA 90041-1777<br>(213) 258-9111 |
| PATRAN II<br>(pre- and post-processor) | PDA Engineering<br>2975 Red Hill Avenue<br>Costa Mesa, CA, 92626<br>(714) 540-8900 |

Category 2: *Finite element programs for personal computers*

| *Program name* | *Program developer* |
|---|---|
| AFEMS | FEM Engineering<br>11222 La Cienega Blvd., Suite 500<br>Inglewood, CA 90304<br>(213) 649-4991 |

Category 2: (*Contd.*)

| Program name | Program developer |
|---|---|
| ALGOR | Algor Interactive Systems, Inc.<br>260 Alpha Drive<br>Pittsburgh, PA 15238<br>(412) 967-2700 |
| ANSYS-PC | Same as ANSYS above |
| CASA/GIFTS/M | CASA/GIFTS, Inc.<br>2761 North Country Club<br>Tucson, AZ 85716<br>(602) 795-3884 |
| COSMOS/M | Structural Research and Analysis Corp.<br>1661 Lincoln Blvd., Suite 200<br>Santa Monica, CA 90404<br>(212) 452-2158 |
| IMAGES | Celestial Software, Inc.<br>125 University Ave.<br>Berkeley, CA 94710<br>(415) 843-0977 |
| mTAB/SAP86 | Structural Analysis, Inc.<br>1701 Directors Blvd., Suite 360<br>Austin, TX 78744<br>(512) 444-0555 |
| MSC/Pal 2 | Same as MSC/Nastran above |
| NISA | Engineering Mechanics and Research Corp.<br>P.O. Box 696<br>Troy, MI 48099<br>(313) 689-0077 |

EGA graphics card and colour monitor. If hard-copy output is needed, some type of printer/plotter is also necessary. If colour slides of model results are desired, an economical approach is to take a time-exposure of the computer monitor, using a 35 mm camera and a tripod. One should be able to purchase a complete computer system including a printer as described above for less than US$5000. By upgrading the previously described 80386 system to include a 150 Mbyte disk, it should be possible to perform any of the simulations presented in Section 4 of this chapter.

System requirements for the Cosmos/M program on an Apple Macintosh II are similar to those described above, with the exception that 8 Mbytes of RAM are required. Anyone interested in acquiring a finite element program should contact the program developers to obtain a detailed list of hardware and operating system requirements. It is often possible to obtain a free or inexpensive demonstration or evaluation version of a particular program. In this way one can try out a program and thereby ensure that the program works with the particular hardware and operating system in question.

The cost of finite element programs varies greatly. Programs having similar capabilities can have markedly dissimilar prices. For any individual program the cost varies depending upon the analysis capabilities one needs and the computer system on which the program will be installed. For example, the cost of the Cosmos/M program ranges from US$1000 for a version capable of linear statics and dynamics running on an IMB PC/XT or AT (US$4500 on a DOS-based 80486 system) up to US$20000 for a full-featured version running on a high-performance desktop workstation (e.g. DECstation, Sun SparcStation, HP 9000/800, Silicon Graphics Personnel Iris). In addition, some finite element programs can be bought outright for a one-time fee, while others are available only through a yearly licence or lease. University and educational versions of many commercial finite element programs are available to qualified users for a fraction of the cost (sometimes less than 5%) of the standard commercial licence fee. One should be aware that educational versions of some programs may have reduced capabilities. The program developers may limit the number of nodes or elements in a model or they may limit the wavefront size which effectively limits the analyses that can be performed to two-dimensional problems only. Some developers also specify that university users can only use the programs for teaching and non-profit research purposes. Additionally, the program developer may provide only limited technical support.

## Acknowledgements

The research presented here was supported by the Department of Veterans Affairs, Rehabilitation Research and Development Service through Merit Review grants A294-2RA and A501-RA. The authors would also like to acknowledge the generous support of Cray Research, Inc., the National Science Foundation, and the San Diego Supercomputer Center. This work is the result of collaborations with the following colleagues: David Fyhrie, Tracy Orr, David Schurman, Robert Whalen, and Marcy Wong.

## References

1. Zienkiewicz, O. C. (1977). *The finite element method*. McGraw-Hill, New York.
2. Huiskes, R. and Chao, E. Y. S. (1983). *J. Biomech.*, **16**, 385.
3. McNeice, G. M., Eng, P., and Amstutz, H. C. (1976). In *Biomechanics V-A* (ed. P. V. Komi), p. 394. University Park Press, Baltimore.

4. Hampton, S. J., Andriacchi, T. P., Galante, J. O., and Belytschko, T. B. (1976). *Proc. 29th Ann. Conf. Engng. Med. Biol.*, p. 321.
5. Carter, D. R. and Hayes, W. C. (1977). *J. Bone Jt. Surg.*, **59A,** 954.
6. Reilly, D. T. and Burstein, A. H. (1975). *J. Biomech.*, **8,** 393.
7. Davy, D. T., Kotzar, G. M., Brown, R. H., Heiple, K. G., Goldberg, V. M., Heiple, K. G., Berilla, J., and Burstein, A. H. (1988). *J. Bone Jt. Surg.*, **70A,** 45
8. Bergmann, G., Rohlmann, A., and Graichen, F. (1990). *Trans. Ortho. Res. Soc.*, **15,** 2.
9. Betts, K. (1990). *Mech. Eng.*, **112,** 59.
10. Carter, D. R., Orr, T. E., Fyhrie, D. P., and Schurman, D. J. (1987). *Clin. Orthop. Rel. Res.*, **219,** 237.
11. Carter, D. R. (1987). *J. Biomech.*, **20,** 1095.
12. Trueta, J. L. (1957). *J. Bone Jt. Surg.*, **39B,** 358.
13. Beaupré, G. S., Orr, T. E., and Carter, D. R. (1990). *J. Ortho. Res.*, **8,** 651.
14. Beaupré, G. S., Orr, T. E., and Carter, D. R. (1990). *J. Ortho. Res.*, **8,** 662.
15. Orr, T. E., Beaupré, G. S., and Carter, D. R. (1990). *First World Congress of Biomechanics*, p. 192. University of California, San Diego, CA.
16. Mackerle, J. (1986). *J. Biomech.*, **24,** 657.

# 8

# Electromyography

CARL GANS

## 1. Why electromyography?

Animals move. Much of their locomotor activity, as well as feeding, drinking, ventilation, defence, and mating involve forces and movements that commonly are generated by muscular action. Hence, studies of the functional morphology of animal behaviours, of the biomechanics of their musculoskeletal systems and of their control characteristics commonly require information about the time and magnitude of muscular action. Electromyography is the premier technique for generating this information. (In the literature, variants of the terms electromyography and electromyogram, are commonly abbreviated as EMG and EMGs. This convenient usage is followed in the present text.)

EMG signals are relatively simple to obtain. However, the body of an animal contains multiple often simultaneous signal sources; hence, these signals may interact. Also, the body is apt to act as a source of electrochemical reactions; thus movements of the tissues, and of electrodes within the tissues, will generate electrical signals by themselves. The multiplicity of overlapping signals requires some skill in selecting equipment, in developing recording techniques, as well as in interpreting the signals. Hence, major sections of this chapter are dedicated to the selection of equipment and techniques to permit one to record signals that can be interpreted unequivocally. It is expected that readers of this chapter are planning to purchase most components 'off the shelf', rather than to construct them. Hence, I emphasize properties relating to their selection and usage, rather than their design and construction.

EMG potentials result from the flow of action currents from muscle fibres through the electrical impedances presented by adjacent tissues. Thus, it is useful to familiarize oneself not only with some aspects of electronic technology, but also with basic concepts of electricity. Review of texts in these fields (1, 2), is well worth the effort. Furthermore, this relatively brief account of EMG provides only an introduction to many other topics involving the recording of muscle activities. For more details, see Basmajian and de Luca (3) for an introduction to human and clinical topics, Loeb and Gans (4) for an introduction to the usage of the technique, practical hints, lists of vendors, and ancillary aspects in animal experimentation, and Knik *et al.* (5) and Stålberg

(6) for useful details on analysis. Reference to simple electronics guides will facilitate the minimal soldering and connector construction required.

Naturally, most students do not record EMG signals to understand the intrinsic nature of the wave forms; rather, they aim to correlate these signals with the mechanical actions effected by the muscles. This correlation requires another suite of techniques and tests, which are introduced here. Consideration of how organisms, consciously and automatically, control their muscles also requires an understanding of activity (or firing) patterns of muscles. Finally, this chapter notes some minimal reporting techniques required to communicate the significant portions of the findings and to assure colleagues that the results are reliable.

Most EMG studies are likely to be performed in living, conscious animals which can feel pain that represents a mechanism which informs them that something is wrong. Because animals tend to change their behaviour in order to avoid this unpleasant stimulus, experimentalists must minimize it in order to gain an understanding of normal movements and normal behaviours. This dictum is intrinsic to functional anatomy and transcends obvious moral and legal questions. EMG by itself may be invasive, but need not cause discomfort or pain. Ancillary techniques, such as motion and force recording (see Chapter 3), must maintain the same standards.

## 2. Principles of electromyography

Active muscle fibres attempt to shorten and in doing so exert tensile forces through their surfaces. They are activated by the chemical transmitter acetylcholine which is released from the motor end-plates of motor neurons in response to the arrival of action potentials. The transmission of electrical signals along both neurons and muscle fibres is mediated by changes in the chemical permeability of their membranes. The resulting currents spread rapidly along the surfaces (and into the depth of muscle fibres via the T-tubule invaginations of the surface membrane). The changing electric fields can then be recorded by electrodes that serve as antennas detecting electrical differences. The differences indicate that the muscle is active.

It is essential to determine first whether a detected signal indeed derives from a muscle. Muscle fibres are generally subdivided into two practical categories, spiking (twitch) and non-spiking (tonic), depending on the rate of change of electrical and mechanical activity. Spiking muscle fibres show a rapid spread of electrical and mechanical activity so that the rise and fall of the signal is seen as a brief spike on an oscilloscope screen. In contrast, non-spiking fibres generate tension more slowly in response to gradually spreading potentials of lower amplitude. Only spiking signals (from phasic fibres) are commonly recorded as EMGs. Analysis of non-spiking fibres requires special and often more complex approaches, mainly designed to filter out extraneous signals without blocking the low-frequency and low-amplitude electrical (tonic) signals.

In vertebrates, the spiking muscle fibres and the motor neurons that co. them are typically assembled into motor units. Each such unit comprises ɛ. single motor neuron that innvervates between 2 and 10000 muscle fibres; within the muscle, the axon of the motor neuron branches repeatedly and each branch forms a potentially activating end-plate on one muscle fibre. With few exceptions, the muscle fibres of adult tetrapods each receive a single innervation of a single motor neuron. In general terms, the activations (mechanical twitch or electrical spike) of a muscle represents a unit; this allows one potentially to count the number of spikes and to estimate the relative magnitude of muscular activation. However, activity does not necessarily correlate with excursion or force. First of all, force is a function of the magnitude and direction of muscular displacement (shortening versus lengthening) and of the velocity of the muscular ends. Repeated stimulation of fibres that have not had the chance to relax will produce increased contractile force, resulting in a tetanus. Trains of spikes may then maintain fibres in the contracted state, either in fused or unfused tetanus. It is assumed that most muscles develop fused tetanic contractions during many functional activities; however, this is not generally true and deserves to be tested in specific cases. Hence the EMG firing rates ordinarily are quasiperiodic, rather than regular.

The electric currents that produce the potential differences being detected by the electrodes arise on the surfaces of individual fibres. The flux or current density decreases with distance from the generating fibre and is additive for multiple fibres. The flux is obviously greater if measured along the length rather than perpendicular to the fibre. Consequently, the signal generated by the fibres of a single motor unit will differ depending on the location and orientation of the recording electrodes. Anatomical information regarding muscle fibre architecture and the distribution of motor units (see following section) obviously will be important. Information about the pick-up distance (reception capacity) of particular electrode configurations indicates the extent to which the muscle is sampled. If the field is too wide, there is risk of combining signals from adjacent motor units or muscles; if it is too narrow, one may only sample from a single motor unit, rather than most motor units in a particular area. The physiological parameters of tetrapod motor units are fairly well characterized; however, it is useful whenever studying a new or possibly unusual species to start by placing several electrodes into close configuration and to compare the signals detected. The area of the electrode tips, their impedance (relative to the tissue being sampled) and the distance between tips will determine their sampling zone.

## 3. Essential anatomy and physiology

Whereas the contractile materials are always variants of actin and myosin, muscular cells have arisen several times and the contractile tissues of different groups may not be homologous (7). The muscles of arthropods and verte-

ly striated, and the mechanics of striated muscles have
atic. Basically, the contractile system consists of parallel
ely of thin and thick filaments, that can slide towards or away
and are each bonded along their middle, like sets of trivets
ate (Z-line) joining the thin filaments is taken as the border
serial units of striation, or sarcomeres, and may be connected
the sarcolemma. Tension and motion are generated by cross-
bridges that can form between defined zones of the filaments; the number of
bridges that can form for a given degree of filament overlap determines the
force generated upon activation (*Figure 1*). Whereas the filaments are shown
in textbooks to lie freely between each other, the tips of the thick filaments
are also connected to the Z-line plates by titan, an elastic material that
accounts for most of the resistance to stretch of passive (inactive) muscle.

In vertebrate muscle, all fibres of any particular motor unit may be characterized as belonging to a distinct type, the types differing in strength, fatigue-
ability, and rate of contraction. Generally, one notes the above distinction
between twitch (spiking or phasic) and slow (non-spiking or tonic) fibres, with
the twitch fibres divided into fast and slow, fatiguing and non-fatiguing.
Although various intermediate types have also been characterized, the
physiological and histological patterns of each fibre type are not always
congruent. In general, tonic muscle fibres are characterized by multiple inner-
vations, with their neurons passing in parallel to them and sending branches
to several sites along the length of the fibre. A few kinds of non-mammalian
twitch fibres also may have such multiple innervation. Unfortunately, the
fibre types differ for the several classes of vertebrates, with distinct categories
being recognized in fishes, amphibians, reptiles, mammals, and birds (10–14).

The force that a muscle exerts depends on a number of factors, making the
assessment of force from EMG problematical. The first factor is the absolute
and relative change of length of the muscle fibres, which in turn depends on
the arrangement of fibres in the muscle (*Figure 2*; see ref. 9). The greater the
amount of contraction and the shortening velocity of the free end, the less
force is produced as the result of activation (*Figure 3*). Consequently,
parallel-fibred muscles will produce a moment (torque) about a fulcrum that
is proportional to their mass and essentially independent of the 'moment arm'
of their placement. Also, stretching of muscles coincident with activation
increases the force exerted. (Such stretching likely results from external
forces or from kinetic energy previously generated and stored as momentum
in other parts of the body.) The correlation between EMG and force is really
quite complicated, as one must also consider the stiffness of the muscle in
response to stretch.

In healthy animals, the fibres of motor units overlap and intercalate, so that
multiple motor units will be noted in any region and assayed by the EMG.
Indeed, it has recently become clear that the fibres of most muscles (in
animals the size of a rat or larger) are shorter than the whole muscle (15). The

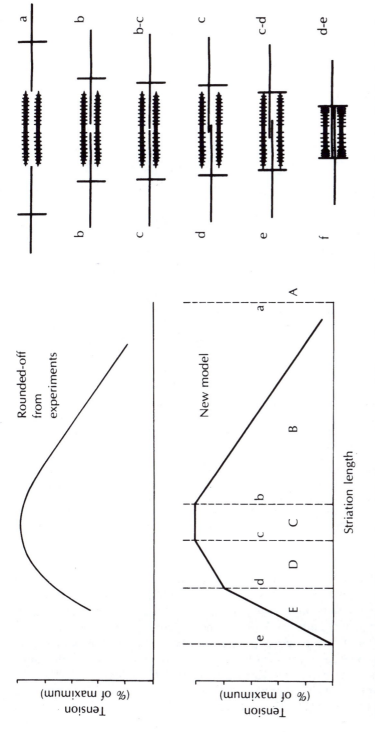

**Figure 1.** The length–tension curve for a complete sarcomere and the overlap pattern of the myofilaments, showing that the degree of overlap between the thick and thin filaments determines the number of cross-bridges that can form and in turn correlates to the force generated, in this case during isometric (non-shortening) stimulation. Patterns of overlap of the myofilaments underlie the nature of force production. *Top,* the length–tension curve represents a smoothed-out generalization based on the actual data obtained for isometric twitches of a muscle. It can be compared to the series of straight lines that describe the actual behaviour of the fibres. *Bottom,* the model is divided into five regions that correspond to the degrees of overlap between the filaments. (From refs. 17 and 18.)

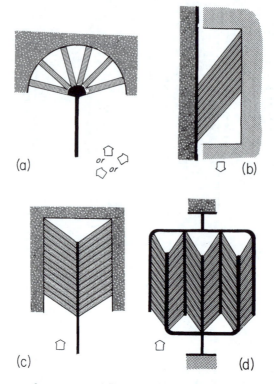

**Figure 2.** The forces of some angled fibres can be resolved into useful components and those that pass in directions which permit them neither to induce movement or force benefiting the animal. Permutations of muscle fibre arrangements. In each case, the force transmitted by the tendon lies at an angle to the force produced by the individual muscle fibres. (a) Radial arrangement of fibres. (b) Singly pinnate muscle. (c) Doubly pinnate mucle. (d) Multiply pinnate muscle. In (a), activation of the different fibres would not only generate differential force, but also shift the site at which the muscles attach to the tendon. The angle of the fibres will change as the fibres shorten. In (b), the surfaces of origin and insertion are maintained in parallel by mechanical stops; in (c) and (d) they are maintained in parallel by the action of fibres inserting on their opposing sides. (After ref. 8.)

assumption that the fibres of all strap muscles are simple cylinders that run end-to-end is wrong. Also, these fibres are spindle-shaped and terminally attenuate; they commonly interdigitate for more than 50% of their length. The fibre arrangement may be characterized by staining the motor end-plates, perhaps by visualizing their cholinesterase. Nerve stains disclose that the muscle fibres commonly are associated in columns, rather than motor units being distributed transversely. This may complicate sampling by EMG.

The activating neurons supplying the fibres staggered within a muscular column must presumably be interconnected. Otherwise, contraction of fibres

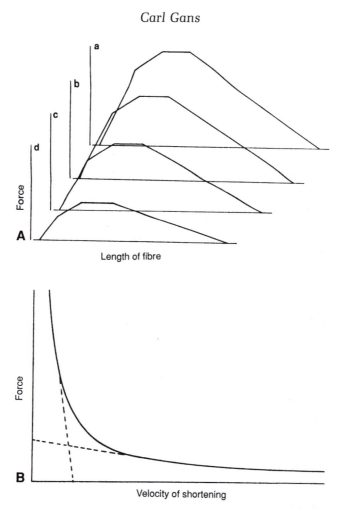

**Figure 3.** The displacement or velocity of the free end of a muscle relates to the force generated. Muscles able to shorten (and achieve high velocity) will reduce the force, whereas those being elongated will increase it. **A**: Set of length–tension curves for a single sarcomere showing the isometric value **a** and three different rates of shortening (concentric contraction); velocity for **b** is less than for **c** and that for **d** less still. The force at lengthening (eccentric contraction) would be greater than that shown in **a**. **B**: Sample force–velocity curve shows that increased velocity produces a disproportionate decrease in the force exerted. Curves are given as indicators of the general force relations and will differ depending upon the muscle tested and the environmental conditions at activation. (**A**: after ref. 9; **B**: after refs. 18 and 8.)

at one level would only stretch the fibres at other levels, without force generation of the muscle as a whole. HRP (horse-radish peroxydase) techniques that 'fill' all processes of damaged neurons (so that these then can be traced, either in serial sections, or by clearing the entire preparation; see Chapter 9) show that some muscles have columnar neuronal architecture;

however, the number of fibres in parallel may differ along the length of the motor unit. In other muscles serially attached groups of fibres are innervated by independent motor units. Consequently, the EMGs of muscles may have to be sampled both at several parallel sites and along their length. Clearly, the neuronal architecture of the entire muscle deserves to be checked in parallel with that of its component muscle fibres.

In various muscles one sees groups of motor units that each are selectively recruited to subserve distinct functions. Sometimes such 'task groups' may be detected by their placement into distinctly oriented fascicles. In other cases the task groups are parallel and intermingled, so that they should produce equivalent skeletal actions. It may be possible that such subdivisions reflect historical effects, perhaps remnants of the past fusion of separate muscles. In other cases, differing muscle task groups may actually exert distinct actions, for instance, by means of forces exerted laterally via the epi- and perimysial coatings, rather than via attachments of the fibres at their ends.

# 4. Recording the EMG

## 4.1 Wiring and amplifier considerations

The electric currents that pass through the tissues of an animal generate potential gradients in that tissue. Potential differences are measured by recording and comparing the value of two local potentials relative to each other and to a relatively stable external value (commonly referred to as ground). (See Sections 4.2.1 and 4.2.2.) Equivalence of the detecting surface of each wire of an electrode, of the connection to the recorder and of the recording circuit itself, makes it possible to eliminate the common portions of the signals by subtraction; this is the common mode rejection of true differential recording fundamental to most EMG recording.

The initial comparison is carried out within a differential preamplifier that establishes the difference, which tends to be on the order of millivolts, between the two recording surfaces. (See Section 4.2.5.) It then amplifies the signal between 100 and 10 000 times, resulting in values of 0.1 to 10 volts. As the wires between animal and recorder carry signals at very low levels, movement of these cables in the different electrical and capacitative fields of the room can easily generate substantial noise. Hence it is useful to pay attention to the configuration of these cables and to keep them stationary and as short as possible. For some purposes, these aims are achieved by mounting the amplifiers on the animal. From here, the signal may then be transmitted either by a long tether or by telemetry. However, the weight of the pre-amplifier circuit and power source may be substantial (and may affect the animal's behaviour), particularly whenever multiple EMG signals must be recorded at one time.

Older preamplifiers tend to produce signals with little power, sufficient only for driving an oscilloscope. The substantial power requirements of chart and tape recorders may induce signal attenuation. Inclusion of power amplifiers assures that change of the recorder configuration will not modify the voltage.

The amplifier output may be observed on the CRT (screen) of a multi-channel oscilloscope. (See Section 4.3.) Oscilloscopes normally have frequency responses that are sufficient to preserve the absolute maximum voltages of the spikes; however, these signals are ephemeral unless recorded on film or other photosensitive media. Chart recorders provide more permanent records, but most have substantial mass (inertia) in their pen circuits; hence they are intrinsically slower and have limited frequency responses. Some solutions to this are covered in Section 4.5.

It is best to save the signal sequence on a magnetic medium, most commonly by recording it on magnetic tape. Inclusion of an FM channel permits storage of relatively slow speed physiological signals reflecting movements of the animal. One may use an A/D (analogue to digital) converter to digitize signals and store them on disk. The tape-recording step should not be omitted; it involves a relatively inexpensive but proven technology and the tape remains available for processing by different means. Visual and auditory inspection and comparison of records also remains a critical and unexcelled test of signal quality and correlation. (See Sections 4.4. and 4.6.)

Once the EMG signals exist in some storage medium, their onset and cut-off, as well as the number and rate of spikes and the magnitude of the signals, can be considered and evaluated relative to mechanical events by means of computer algorithms to analyse the signal statistically. (See Section 4.7.)

## 4.2 Electrodes: design and placement

### 4.2.1 Principles

The selection of electrode materials, their configuration, and their placement must be the first and critical step of any EMG analysis. These considerations depend primarily on the absolute size, configuration, and position of the muscles from which recording is to proceed.

The materials of the electrode should be relatively stable, and should avoid deterioration particularly of the kind that poisons the tissues. Various noble metals, such as platinum and gold, though relatively expensive, are suitable for electrode design. Because of its softness, gold is prone to fatigue over longer periods (days to weeks) of recording. However, the application of gold as a coating to harder metals or plastics, makes it suitable for prolonged implantation. In contrast, silver and particularly copper, tend to electrolyse and form relatively toxic solutions that may gradually kill the surrounding cells. Several kinds of stainless steel wire, 0.015-inch to 0.001-inch in diameter have recently been used with success; they permit long-term implantation without deterioration. Some thicknesses are available with biocompatible dielectric coatings, such as Teflon.

The electrodes should provide a pair of lead off surfaces (transmissive contact zones between electrode and tissue) of equal area. Their geometrical equivalence with respect to the muscle fibre will facilitate differential amplification and is more important than the area of the exposed surfaces or length of leads. The electrodes should be discriminatory, detecting the signals generated in a small volume of the muscle (see below). In order to avoid spurious signals, one may connect the animal to a common ground (independent of the electrodes) and perform the recordings in an area that is shielded against stray electromagnetic effects (see Section 4.4).

The surface of each electrode should be as close as possible to the tissues generating the signal. As the depolarization wave passes down the fibre, pairs of electrodes oriented along it length will each encounter a change of the electric field; hence, the differential recording will appear as an antenna yielding an excellent indication of the spiking event. In contrast, a pair of electrodes located equidistant from the sides of such a fibre (with the line connecting the electrode tips lying perpendicular to the fibre) will encounter signals of equivalent magnitude; these will cancel out in differential recording. This makes it necessary to compare the responses of an electrode pair oriented so that the bared tips lie along a line parallel to the fibre (*Figure 4*). As fibre angles and positions differ throughout muscles, the optimum placement of electrodes may change, even during contraction. Compromise placements may make the responses less than optimally repeatable. However, orientation remains critical. It affects not only the appearance of the signal, but also its absolute magnitude.

## 4.2.2 Electrode designs

The basic dimensions of EMG electrodes should be as follows; for additional details see ref. 4:

(a) Bipolar separation of tips about equal to the dipole separation of current sources and sinks to be recorded (that is, wavelength of the action

---

**Figure 4.** Diagrams to show passage of an activation wave along a fibre and the ideal placement of electrodes to obtain maximal signal level. **A**: the flow of ions induced in the extracellular fluids around a muscle fibre generating an action potential decreases rapidly with distance from the surface of the fibre. The effect of distance is relative to the dipole separation of the current source and sink, shown here as a single equivalent point produced by the action potential coursing along the fibre. **B**: Two types of bipolar hook electrodes suitable for percutaneous insertion. Note that the simple double hook (*top*) tends to create a bipolar axis that is perpendicular to the insertion angle and that may line up parallel to the muscle fibre dipoles (as shown) or perpendicular to them if the insertion tool is turned 90 degrees on the axis shown by the arrow (undesirable). The offset twist hook (*bottom*) tends to produce a bipolar axis that parallels the angle of insertion, and it is thus most suitable for muscles that may be approached tangentially (or that are very pinnate). (After ref. 4.)

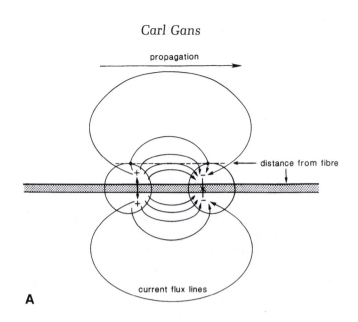

propagation

distance from fibre

current flux lines

**A**

Simple Double Hook

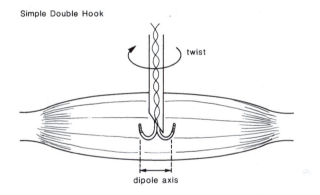

twist

dipole axis

Offset Twist Hook

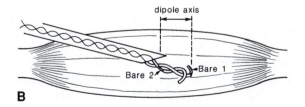

dipole axis

Bare 2

Bare 1

**B**

potential equal to the duration of the events times the velocity of conduction; about 2–10 mm for mammals).

(b) Recording contacts as similar as possible.

(c) Each recording contact as large as possible (one linear dimension about half the bipolar separation).

(d) Keep the electrode pair within one bipolar separation of the signal source. (Signals originating at more than four times this distance will be recorded similarly on each contact and therefore selectively rejected by differential amplification.)

The fixation of the electrode must depend on the nature of the animal being studied. If the animal possesses a stiff exoskeleton, or the recording must be non-invasive, the electrodes have to be placed on the surface, preferably as close to the muscles to be sampled as possible. For non-invasive (human) recordings, skin surface electrodes are available. They are advantageous as surgery always involves potential trauma.

All other electrodes are intrinsically invasive, although they may be small enough to be inserted by methods that are essentially non-traumatic. Needle electrodes, consisting of tiny tubular glass or metallic rods, are suitable only for deeply anaesthetized organisms, as any movement will shift them and likely cause trauma. For that matter, they are likely to be moved by shift of more superficial muscles; this may cause electrical movement artefacts as the metal slides past the tissues.

### i. Surface electrodes

The recording properties of surface electrodes are defined and modified by the dielectric characteristics of the integument. Because these electrodes are always some distance from the muscles being sampled, they are relatively non-selective and are best used over large masses of muscle that tend to act in unison. They do not provide information about the action of individual motor units (unless the muscle is being activated—perhaps voluntarily—at very low levels), and are particularly ineffective whenever regions of the muscle differ in their activity.

Surface electrodes are normally configured in monopolar or bipolar conditions. The silver and silver chloride contacts lie in cups that are filled with a conductive jelly. In order to reduce electrical noise and motion artefact, the contacts tend to be adpressed to the skin by pressure contacts or suction cups. The external constraints establishing tight contact may pose problems in animal applications; however, for human studies one notes various special applications (3). Among these are recording for psychophysiological analyses and myoelectric control, in which EMG signals are used to activate prosthetic devices.

### ii. Fine-wire electrodes

These are most commonly used at this time. (See Section 4.2.3.) They may be placed stereotactically into muscles, being oriented by external triangulation or the use of bony landmarks. Commonly, they are inserted by an hypodermic needle, which may then be withdrawn, leaving the wires in place. Prior to insertion, the lead-off surfaces are prepared by removing the insulation from part of the surface of the wire. Also, the pair of wires is generally interconnected mechanically so that the two surfaces retain their orientation relative to the muscle and to each other.

### iii. Patch electrodes

These require invasive surgery, but their use can obtain high-quality, well-isolated signals in anatomically difficult situations. (See Section 4.2.4.) The electrodes are formed on one surface of a dielectric membrane, such as a layer of silastic artificial dura, which restricts current flow. This side is held against the surface of a muscle by surgical glue or sutures. The dielectric properties of the patch restrict the flow of action currents, so that the electrode pair mainly records signals originating from its side of the patch. Consequently, the method is ideal for recording selectively from small or thin muscles lying adjacent to large and stout ones that may be active nearly simultaneously.

Several electrodes may be placed on to the two sides of a single patch and this then located between two muscles, thus recording differentially from them. This lets the patch electrode discriminate effectively between the signals derived from two different, but closely adjacent sources.

### iv. Multipolar electrode arrays

There are various combinations of multipolar electrode arrays. Examples are triple lead-off surfaces, the signals from which are sequentially recorded as pairs. In such arrangements, the output of any single motor unit is likely to display a characteristic wave form on each of the combinations that may permit its identification and localization. Unfortunately, comparison requires that the recordings will be sequential and that activation of the motor units be equivalent. Unless one is dealing with highly trained animals, such methods are useful mainly if it is possible to activate the muscles by stimulating their motor neurons or higher centres of the central nervous system.

### 4.2.3 Fine-wire electrodes

Fine-wire electrodes are the method of choice for many small animal experiments, as they may often be inserted stereotactically with minimal surgery; however, the orientation of their tines often tends to be insufficiently controlled. Electrode pairs are normally composed of single or twisted multi-stranded stainless steel wire. Each such wire runs continuously from the contact area through the integument to an external connector that transmits the signals to an amplification circuit. This seamless construction avoids the potential electrolytic activity that tends to occur whenever different metals are joined within a conductive aqueous solution.

There are multiple insulation types commercially available and those impose different tasks in exposing the contact surfaces. Varnish and enamel coatings may be burned off with low heat, using paper matches. The merit of low heat is that it avoids oxidizing the metal. These problems are reduced for Teflon-coated wires which tend to be stripped more easily. Teflon deforms with minimal heating and may be trimmed to expose a desired length of wire.

Repeatable EMG recording demands that the tips be placed into a predictable position, that they maintain their relative position and that they do not short out, either during insertion or *in situ*. Positional stability is often enhanced by twisting the wires against each other. The electrode pair can be formed of a loop that is hooked over a gimbal (or a hook placed into an old-fashioned hand-drill); this is rotated slowly, while the wires are tensed. The loop is cut asymmetrically, and each free end is bent in a different direction and stripped of its insulation. A drop of glue then fixes the tips relative to each other and the long ends of the wires are threaded through a hypodermic needle (No. 25 for 0.002-inch wire), with the electrode tines bent backward.

To insert the electrode, the skin is nicked with a scalpel tip (to avoid deformation of the electrode), the needle is inserted until its tip reaches the final position in the muscle, and is then withdrawn, leaving the long ends emergent from the skin. The recurved tips normally hook into place within the muscle and local scar tissue forms over the next few days, helping to anchor them into place. Light pressure applied with a finger tip to the site at which the tips lie also helps to avoid slippage as the needle is withdrawn. The path of the emerging wires is important; it must be planned to avoid major blood vessels and nerves. Layers of other muscles are critical as their movements might tend to displace the electrode and stress its attachments within the muscle. Looping the electrode wire between the insertion site and the integument provides slack and minimizes movement artefact. Remember that even such simple twisted electrodes must have their tips lying on a line aligned parallel to the fibre.

The hypodermic needles to be used for insertion should be checked so that there are no constrictions, or internal lips that might catch the wires during threading or when pulling them out after insertion of the wires. The bevel of the tip should be selected to be short, but the tip should be sharp enough so that the needle slices rather than tears through the tissue. The internal edge of the needle may tend to cut through the insulation of the wires during insertion, particularly in penetration of tough skin. This may be avoided by passing the electrode-bearing needle through pre-cut nicks in the integument, by chamfering the internal edge with a miniature hone and by adding a drop of fingernail polish or other hard-drying glue to the bend.

Simple twist electrodes may shorten out if their tines contact each other. Offset twist electrodes compensate for this (*Figure 5*). In these the twisting action is stopped before the anterior portion has been fully combined, and one of the two bare surfaces is arranged along the length of one wire. The twisting then continues, so that one-half the portion of the wire containing the

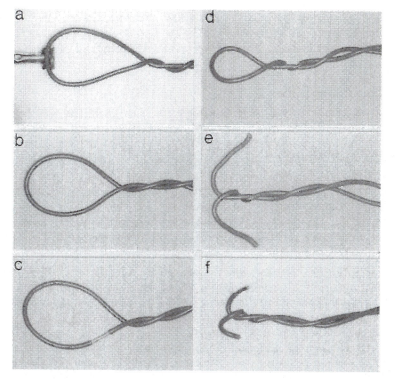

**Figure 5.** Construction of fine-wire offset twist electrodes. Heavy wire is used to model the constructional principles. **a**: A loop of the wire is twisted. **B**: The wire is removed from the hook. **c**: A portion (1 mm) of the insulation is stripped off one side. **d**: The loop is replaced on the hook and twisted until the bared area is incorporated in the twisted portion, with twisting continued for at least an equal length. **e**: The loop is removed from the hook, the tip is split and the tines are bent backward. **f**: A (1-mm) length of one of the tines is bared, thus generating the second contact surface. (From ref. 4.)

first bared wire is included in the twisted portion. Once a zone of twist as long as one-half the bared area has been formed, the remaining bared area of the first wire is recurved and a portion of the second wire is bared. Even if this tine is bent toward the twisted portion, it should not be able to contact the other bare area and short out. As the two lead-off areas are displaced along the axis of the electrode, the offset twist electrode must be inserted more or less in parallel to the fibre axis with the free tine placed more or less along the insertion axis rather than lateral to it.

Whereas twisted electrodes remain the favoured design, for long-term insertion, particularly for use in large and active animals, various foam epoxies bonded to the wire insulation can help to reinforce the electrode. This ensures that the two wires remain in a fixed position during insertion and later recording. However, the epoxy must be allowed to cure before insertion, and

**189**

one must check that none of the fillers and additives are non-toxic. These standards apply to expensive medical grade adhesives; however, the much cheaper commercial grade materials are perfectly usable if these precautions are observed.

### 4.2.4 Patch electrodes

Patch electrodes are highly application specific. The patch (sheet of dielectric material) has to fit the free surface of the muscle. Also, it must be trimmed, so that it may be fixed surgically to fascial regions yet continue to contact the muscular surface during contraction. The latter aim may be achieved by placing the patch between two muscles; it is also possible to form the sheet into a sleeve (or to utilize a plastic tube) and to place the tube around a strap muscle with the electrodes on the inside.

Artificial dura represents an excellent patch material; however, many kinds of synthetic membrane or sheeting may be used. The mechanical properties can be enhanced by coating the polymers, such as silicone rubber, over a reinforcing layer of Dacron mesh. Lead-off areas may be constructed by threading two thin wires through the patch and removing insulation from those portions intended to lie in contact with the surface of the muscle (*Figure 6*). The distance between the wires establishes the recording zone. It is also possible to place a series of contact surfaces along the length of a patch and to record from these in different configurations. Two-sided electrodes are formed in a similar fashion.

### 4.2.5 Electrode connections

The voltage detected by each electrode pair must be transmitted to the preamplifiers. Generally, multiplexing is not used, and a single preamplifier is dedicated to each electrode pair. At the outset, it is important to ensure that the several sets of fine wires are properly identified; this should involve meticulous record keeping and a marking regime. The ends of the wires emerging from the animal may be colour-coded, perhaps with fingernail polish, or threaded through combinations of differently coloured tiny glass embroidery beads, the so-called hishi beads. Self-sticking coloured dots may facilitate temporary coding.

The electrical potentials in the muscle are in the relatively low range of 100 to 2000 microvolts. As noted earlier, the cables leading to the differential preamplifiers are sensitive to electromagnetic interference, as well as to changes in their capacitance. The shorter the distance between lead off surface and preamplifier, the lower the risk of artefact. Hence, it is sometimes useful to place a preamplifier on an animal; this is one of the supplementary merits of telemetry circuits, which normally involve a first stage preamplification prior to transmission. Even if no miniature preamplifiers are available, some other precautions may help. The major one is to avoid potential ground-current 'loops'. For example, the shields or reference electrodes of

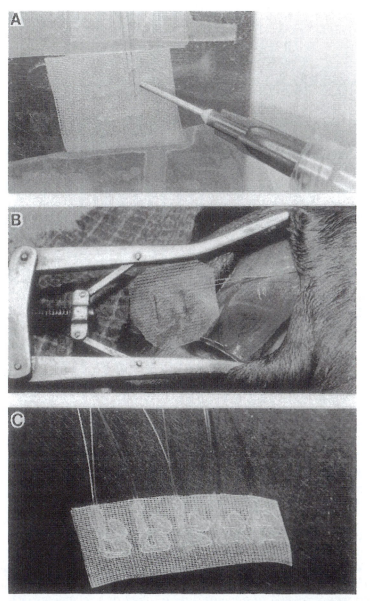

**Figure 6.** Stages during manufacture of a patch electrode. **A**: View from the back shows silicone rubber RTV being applied to the entrance and exit points of the leads to prevent strain relief and insulate any exposed wire. **B**: Recording surface side of a patch electrode, which will be flipped over and sutured at the corners on to the surgically exposed muscle. **C**: Multicontact patch showing five evenly spaced bipolar recording configurations used to compare recruitment across different parts of a broad, sheet-like muscle. (From ref. 4.)

the several cables should be connected only to the preamplifier side, with only one of them attached where the electrode cable reaches the animal. Similarly, the reference electrodes should not contact the shielding of the Faraday cage or the water in which an experimental animal may be swimming. (If the shielding of several cables is connected simultaneously both to the animal and the amplifier ground, currents will likely be generated, particularly if the cables move during recording sessions.)

The array should be designed so that the experimentalist is unlikely to touch the cables during recording sessions. Use of low-noise cables in which the shielding does not move significantly relative to the signal transmitting leads is also favoured. Finally, it remains critical that the cables transmitting the differential signal be equivalent so that signal artefacts show up in each and may be rejected. Hence, use twisted-pair leads; never use coaxial cable with the core carrying one electrode pole and the shield the other.

Animals are commonly allowed to recover for some hours or even days prior to the start of major recording sessions. During this interval the free ends of the electrodes may be held (tied or taped) loosely to the animal's back. However, it is better to attach all of the electrode wires immediately to one unit of a multi-pin connector, as this permits repeated monitoring of the impedance of the inserted electrodes during recovery (see Section 4.4 below). The matching connector should then lead to the preamplification stage by a relatively loose tether that permits the animal to move freely without straining the connection. Various rotating tethers can maintain contact even if the animal turns repeatedly. Most important is that the connectors and their associated cables not be stressed. This is achieved most easily by paralleling them with a tension-resisting connection that is shorter than those conducting signals.

Whereas well-designed electrodes will function with minimal noise in the absence of major shielding, the inclusion of a Faraday cage limits the potential effect of electric and electromagnetic sources of noise. Such a cage should be constructed of metal or metal mesh, and these should form all six sides of the cube, including its floor. The metal walls should be electrically interconnected, ideally by low-resistance braided metal straps. Similar straps should also interconnect the walls and the access doors. Whenever one is recording from aquatic animals, the water can serve as the shielding material. In this case, shielding is needed only for the connections from the tank to the preamplification circuits. All shielding and ground cables should be connected independently to a major sink to establish a reliable ground connection.

## 4.2.6 Telemetry

Instead of transmitting the EMG (and ancillary) signals via cables, they may be telemetered, either by radio or sonar circuits. Sonar has rarely been used, in part because it works best whenever the transmitter and the antenna are immersed in the same body of water. However, radiotelemetry has been used and is now increasing in applicability, both for experiments in the laboratory and for monitoring animals ranging over restricted ranges in the field. Telemetry does have the enormous advantage that the animals may be freed from external restraint and allowed to travel freely. This limits potential for behavioural artefacts, for instance in locomotion.

The limitations of telemetry are that the animal has to carry the pre-amplifier, transmitter(s), and associated batteries, which may be bulky and add substantially to its mass. The battery capacity required increases with the distance over which signals are to be transmitted and with the duration for which the circuit is to be active. Furthermore, the frequency bands available for transmission of biological signals is legally limited (in the US by the Federal Communications Commission), as such transmitters can interfere with commercial channels. Finally, the limited frequency and useful amplitude range of any channel favours use of FM (frequency modulation) and multiplexing for signal transmission.

Numerous companies produce medium-sized multi-channel systems about the size of one or two packs of cigarettes (but weighing 30 to 50 g); these are designed for use on human subjects and animals the size of dogs and cats. (Their price is correspondingly high.) Commonly, the bandwidth of telemetered signals will be truncated, reducing the maximum and minimum frequencies, in order to increase the number of channels that can be transmitted simultaneously. Recently, there has been the advent of a different approach in which each electrode signal is transmitted independently, on a different wavelength and by a separate transmitter. Unfortunately, each channel also requires an independent battery. However, battery life can be extended by use of magnetic on/off switches. Thus, such units can be implanted subdermally and operated by remote control. The system is available in kit form at a modest cost.

## 4.3 Signal treatment

The first treatment of the signal is its extraction from the background activity by the differential preamplifier, that also multiplies the voltage by a factor of 100, 1000, or 10 000 times the differential detected between the two lead-off surfaces. (For any stage of amplification, it is necessary to calibrate using a known input signal. The gain values elegantly silk screened next to dials often represent only a first approximation to reality.) Modern amplifiers simultaneously boost the signal power, reducing the susceptibility to extraneous noise sources.

The voltage emitted by such a preamplifier often includes an offset value that can be modified by set screw or dial. Most display devices will indicate the voltage and often show only a small portion of the voltage. For these reasons the offset voltage may pose problems in amplification. Thus, if one is interested in the rate of voltage increase and fall, for instance, between 0.5 and 0.6 volts, it would be best to shift the offset voltage to 0.5 or even 0.55 volts. In this case a voltage amplification factor of 10 will multiply the range of the signal measuring from either of the new baselines, i.e. in the first case 0 volts will represent the 0.5-volt level and 1.0 volt the 0.6-volt level, with the signal oscillating between these. Should the offset voltage of the initial signal have been 1 volt, it too would have been amplified, probably causing the signal to go 'off scale' or outside the dynamic range of the recorders. Some display systems incorporate procedures that allow reorientation of the signal, so that the trace of the amplified signal remains in the centre of the screen.

## 4.4 Noise and electrode reliability

The signal-to-noise ratio is a critical aspect of any experimental analysis. Ideally, a signal has a straight (clean) baseline and all deflections therefrom have experimental significance. Any departure from this state requires explanation and complicates decisions about the biological events involved. The importance of this process to the experimentalist has led to major efforts at recognizing and avoiding signals from sources that may be very interesting of themselves but are extraneous to the experiment. The three kinds of noise that occur and need to be recognized are:

• wide band (wide frequency range) noise
• narrow band noise
• biological noise

### 4.4.1 Wide band noise

The wide band or thermal noise relates to intrinsic physical properties of the system and materials. It is inversely proportional to temperature, band width, and electrode impedance and should not be a real problem for well-designed differential EMG electrodes.

### 4.4.2 Narrow band noise

Elimination of narrow band noise normally requires identification of its source. The noise source that is most ubiquitous and in some ways easiest to control is that resulting from the frequency of the alternating current providing power to equipment and lights; this is normally near 50 or 60 Hz in Europe and North America respectively. Like other kinds of noise, it can be avoided or compensated for by shielding the animal, adequately grounding the shielding, and ensuring that the cables carrying the signals from the animals do not contain the potential for current loops. Some amplifiers also incorporate notch filters that selectively eliminate any signal at 50 or 60 Hz.

The same approach also applies to various other signals considered to be noise. However, the best way to remove noise is to avoid picking it up in the first place. To do so, the power cables to the several pieces of equipment should be run from the periphery inward, away from the experimental array. Also, cables should be shielded and be stationary, supported rather than being suspended across open spaces in which the experimentalists will have to move. Swinging electrode cables may induce currents as they shift through the inevitable magnetic fields, and movement of the experimentalist among them will most likely cause gradual changes of the interfering fields.

Baseline shifts and gradual (low frequency) changes of signal will not be seen whenever oscilloscopes or similar devices are set for AC coupling. However, FM tape-recorders generally preserve all of the signal, including low frequencies and baseline drifts which may complicate subsequent analysis. These frequencies are best eliminated by passing the signal through a high-pass filter (often just a capacitor) which preferentially transmits the higher frequencies.

Many noise sources are highly directional; this is true for stroboscopic illuminators. Their effect may be reduced or eliminated by placing well-grounded wire grids between them and the animal area. Experiments should always be started with a test of the recording arrangement, substituting a saline-soaked sponge for the animal. If the noise level is more than 10% the magnitude of the EMG spikes, one must start to search for sources. A decision-tree (*Figure 7*) will make the process more efficient, but often it comes down to disconnecting equipment items one after the other.

### 4.4.3 Biological noise

Biological noise derives from other organs and tissues, which often produce signals of equivalent and even greater magnitude than the muscle being studied. The key here is to record the muscle being sampled and eliminate the remainder. Fine wire and patch electrodes do this effectively, particularly whenever their cabling and amplification stages are appropriate. Surface electrodes are much less specific.

It is always important to keep the electrode source impedance low, relative to the amplifier input impedance. Consequently, the actual impedance of the electrodes, pair-wise and for each individual lead, should be checked shortly after implantation, and regularly thereafter. Shorted electrodes and broken leads will rapidly be discovered in this way. At the same time, one should develop an expectation of the voltage levels that the muscles are likely to display to the electrodes of the characteristically spiky nature of EMGs. Whenever the signals are too low or too high, one should suspect motion artefact. Quick differentiation between the EMG's crisp sounds and continuous 'groaning' sounds of motion artefact can be facilitated by the use of a simple loudspeaker circuit connected to the output of the preamplifier.

### 4.4.4 Cross-talk

Cross-talk among circuits is easily checked for. Cross-talk must be suspected, and its potential occurrence eliminated, whenever the signals deriving from

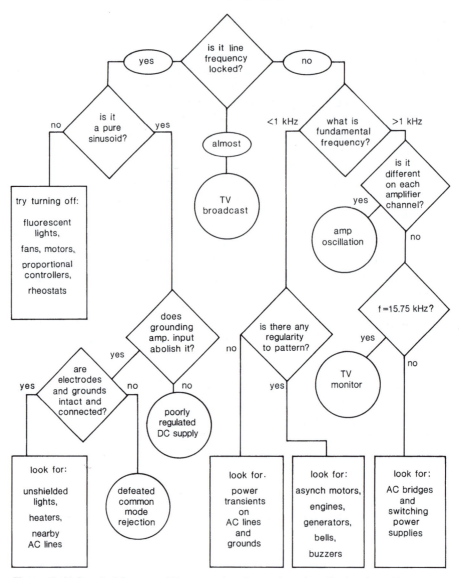

**Figure 7**. Noise decision-tree. The cascade of questions is effective in asking about interfering noise, to converge quickly on the likely source of the problem. (From ref. 4.)

two distinct electrodes regularly appear to parallel each other. Such cross-talk can, of course, occur in several places. It is unlikely to be a problem for modern signal processing and recording equipment, unless the units are inappropriately interconnected, for instance in multi-channel telemetry. Cross-talk among cables is more common, particularly whenever the cables

**196**

have the potential of ground loops. Cross-talk among closely adjacent electrodes may occur whenever the muscles emit signals of markedly different amplitudes. Three-dimensional recording, using patch electrodes and the insertion of simple barrier patches, can often determine whether such signals indeed derive from adjacent muscles, rather than those into which the electrodes were placed. Deinnervation and selective temporary nerve and muscle blockage provide further tests. Reversible manipulations (blocks) are best (although sometimes impractical). Remember that the voltages represent vectors, and that reorientation of the antennas represented by the electrodes is likely to have a major effect on the amplitude of the signal, indicating its source. In general, the issue of signal reliability requires considerations equivalent to those posed by correlation of the EMG with mechanical events and is discussed further in Section 4.7.

## 4.5 Signal storage

The next problem to consider is the display and storage of the information provided by the multiple electrodes. All display systems raise two questions—namely, the frequency response of the signal recorded, and the permanency of the record. Simplest are oscilloscopes, which can show several signals simultaneously and store them briefly. By their nature, such displays provide very fast, but ephemeral responses. Storage oscilloscope display several fancy and often expensive variants. The displayed image can be recorded (as hard copy) on film (or videotape), but such recording is indirect and further processing of signals stored in this way is complex. Also, the display speed is likely to be fixed to that of the initial recording.

Chart recorders provide an immediate and very permanent hard-copy by tracing the signals on paper; the records may later be compared, analysed, and displayed in publications. Most chart recorders are slow, and have poor frequency responses; however, these may be improved by adding servo motors which apply supplemental force as the limits of the pen excursion are reached. They may incorporate complex linkages that provide a rectilinear rather than a curvilinear record. Some recorders reduce the mass of the 'writing' element to increase frequency response. The record is furnished by ink sprayed through suspended capillaries or by light beams directed at sensitized paper. A further improvement of the frequency response involves use of variable-speed tape-recorders. Signals recorded at a fast tape speed are written while the tape plays them back more slowly (be sure that the electronics permit this without degradation).

Whereas a 'hard' copy of the signals is desirable, it is difficult to analyse. Signal sequences are better saved for future analysis on a magnetic medium. The most cost-effective storage uses magnetic media, such as analogue or digital tape. Tape is remarkably cheap, so one can afford to record long-term experiments, later engaging in statistical analyses of repetitive sequences and looking for the rare event. Also, we know much about the stability of tape; it has long been used. Recording may be with multi-channel AM or FM systems. The latter tend to be more expensive but can store data over a wider frequency range, such as the output of ancillary (relatively slow speed) physiological signals might be correlated with the EMGs. Thus the output of transducers that indicate movements of animals may be stored in parallel with simultaneously monitored EMG signals. Simple systems may use commercial quadraphonic tape systems, the frequency response range (50–12 000 Hz) of which are more than adequate for most EMGs.

All of the signals may be recorded routinely on tape, deferring decisions and ultimate analyses to a later time. Tape systems are usefully combined with a simple chart recorder. Both are maintained in recording mode during the experiments; however, the tape will record at its regular speed and the paper of the chart recorder advance more slowly. The paper tracing may be marked with a hand-written experimental protocol reporting animal activity and also preamplifier, amplifier, and recorder settings. The record then provides a permanent guide to those portions of the tape that should later be analysed or printed in an expanded version, either off the chart recorder or by photographing an oscilloscope screen.

Various microcomputer configurations (for instance, simple add-in boards), can acquire multiple signals directly; by analogue to digital (A/D) transformation sometimes fast enough to proceed on line and even to be stored on disk. However, EMG analysis requires rather high sampling rates (at least 4 to 6 times the upper frequency limit of the signals) and consequently has substantial storage requirements. Software packages can reconstitute all or selected portions of parallel traces on screen. Some programs generate such graphical presentations and others perform digital analyses, providing tables of numbers subject to statistical manipulation. One may test the similarities and differences among activities performed by organisms, in age-associated or other aspects. Consequently, the analysis of variability has been greatly simplified and every postal delivery seemingly offers new technical products with additional features.

However, it seems useful to caution that the traces generated even by the best of these programs are very far from raw data. This suggests resisting the temptation to omit the tape-recording step. Tape is relatively inexpensive, provides a proven medium for relatively long-term storage, and leaves the data available for processing by different means. Possible signal degradation by the electronic circuitry can then be checked for, as can the effect of inappropriate sampling rates. Also there remains merit to an oscilloscope for routine check of the reliability of the digitization. Remember that computer systems have been known to fail, and so have their storage media.

Once the EMG signals exist in some storage medium, they may be evaluated as hard copy, or analysed statistically by use of computer algorithms.

## 4.6 What parameters to sample?

Some of the spikes and wiggles of an EMG seen on the screen are irrelevant; however, others may be highly significant. How does one decide what and how to sample data traces?

The first question concerns the sampling frequency—that is, the number of times a trace must be sampled per unit time to permit acceptable reconstruction. The theory behind sampling rate is simple to visualize. Assuming that a spike takes a unit time from start of rise to return to baseline, one can see that there must be a minimum rate of samples taken to ensure that at least one sample will occur at or near the peak and a greater number so that both the rise and the fall of voltage are sampled. As the sampling frequency decreases relative to the time of spike rise and fall, there is a significant chance that the spike tip will not be encountered. Sampling of the rising or the falling slope may provide the highest observed voltage and the spike height will be randomly underestimated, a phenomenon called 'aliasing' (see Chapter 3 as well).

As EMG signals are only quasiperiodic, their signal trains incorporate a wide range of frequencies. Fourier analysis, which is available on many software packages may establish the distribution of frequencies within a signal; it may differentiate between signals provided by tonic and phasic fibres. Only some of this bandwidth may be related to the particular physical events being studied. This means that the sampling frequency may be kept at a level in which some of the signal frequencies cannot be sampled, but only if these higher frequencies are filtered out before sampling. For instance, it has been suggested that certain vertebrate EMGs may be adequately reproduced by sampling frequencies no higher than 4000 samples per second (13), a value that may reflect the rise time of the spikes as much as spike repetition frequency. For a given sampling rate, frequencies above one-sixth of this rate should be removed by first passing the signal through a low-pass filter (set for that frequency). This avoids aliasing.

Various kinds of information in the EMG may be of importance and may be usefully quantified. Obvious among these is the duration of the EMG and the start and cessation of its activity, often relative to key physical events (mechanical and kinematic). The kind of onset is also important as it may be characteristic of particular actions. In some instances one sees a few anticipatory spikes before the onset of the major EMG burst, seemingly by simultaneous action of many units; in other instances, onset and cessation involve a discrete shift between zero and full magnitude. This difference is not always explainable but deserves mention.

The magnitude of the EMG (and its changes during the major activity period) is often important and has routinely been correlated with the forces and excursions produced by the muscular actions. It may be determined in many ways. Often the signal is full-wave rectified (inverting the negative portions of spikes to positive). The numbers of spikes, their individual magnitudes, and the number of positive zero crossings (indicators of coincident spikes) can then be counted. Some investigators have argued that there is little reason to determine such variants separately; however, the descriptors often vary independently, suggesting that they reflect different apects of the myoelectric activity.

Summations of the spike amplitudes, or of the spike numbers (or zero crossings) multiplied by the mean spike height represent two logical approaches for assessing the energy level in the signal. A simpler method involves tracing (mechanically or electrically) an envelope over the heads of the highest spikes seen; this approach is reminiscent of the classical method of trimming out the burst marked on chart paper and planimetering the area or weighing the paper. Electronic integration circuits operating on the rectified signal can accomplish the same result, producing a moving average over time. The properties of the signal integrating system must be understood and reported.

## 4.7 Correlation of the EMG

Muscles obviously act to induce movements. However, making the correlation between the EMG and a particular movement requires certain specific tests (*Figure 8*). The first consideration here is that of excitation–contraction (E/C) coupling. The EMG signals appear at the time that the excitation starts to spread out from the motor end-plates. Hence, it differs from the signals discussed above in that it predicts the future generation of force or excursion rather than responding to it. There is a finite interval between the EMG and the beginning of any mechanical event and a greater interval until maximal force has been generated. Similarly, cessation of the EMG is only later followed by change of the mechanical action. The E/C interval is temperature-dependent and species-specific; it is often stated to last 20 msec. However, it may be much shorter. The coupling interval needs to be established and taken into consideration in testing for correlation between EMG activity and mechanical action.

Next is the question whether particular EMG patterns are regularly associated with the behaviour; in short, is there substantial cross-correlation? Also important is whether the EMG signatures of particular muscles are necessary and sufficient for performance of the action. Whenever the actions are complex, the movements are likely to be functionally partitioned. Different muscles or muscular compartments must be expected to be involved in slow and fast actions, in eccentric and concentric force applications and in ones involving forceful versus weak movements. This raises questions about which muscles do generate the movement. EMGs should be recorded simultaneously from all muscles that could be associated with the action. The tests may well become complex, as muscular actions are often modulated by sensory components.

Intact, normally behaving animals rarely allow the movements and active

## Carl Gans

Is the behaviour cyclic?

Yes?                                                                No?

Establish variability among cycles.

Test all muscles possibly associated.
Are they active: NEVER? SOMETIMES? ALWAYS?

If never, ignore.
If sometimes, reconsider.
If always, establish:

Intervals:
   ONSET of EMG to onset ACTIVITY.
   ONSET of EMG to maximum ACTIVITY.
   PEAK of EMG to peak ACTIVITY.
   PEAK of EMG to cessation ACTIVITY.
   CESSATION of EMG to cessation ACTIVITY.

Consider excitation-contraction coupling.
Does pattern of EMG activity correlate with
   that of force (velocity)?

Can mechanical activity proceed without muscle?
Can activity proceed in absence of synergistic
   muscles?
Are any antagonists active simultaneously?

**Figure 8.** Decision-tree for establishing the contribution of a muscle to a particular function or set of functions. Naturally one will also consider the extent to which behaviour is characteristic of the individual, the population, the species or some supraspecific group.

forces of their muscles to be monitored directly. Instead, the experimentalist tends to record their actions indirectly by noting various actions, generally of the animal as a whole. Any aspect that is recorded in purely behavioural study, such as sound, displacement (in locomotion, feeding, grooming or mating), or pressure (or force), may be used to estimate the actions of muscles. Similarly, such physiological parameters as blood flow, air flow, and turgor are of interest in some studies. Consequently, one obtains physical records on cine-film and video tape, on audio tape, and as the output of physiological transducers. Audio and physiological records are obviously the easiest to analyse, as they can be stored on tape in parallel with the EMGs. This facilitates correlation. Analysis of films and videotapes is covered in Chapter 3.

Certain correlative techniques deserve attention. Electric stimulation was mentioned above. (See Section 4.4.) This must involve large, low potential electrodes made of non-corrosible and non-toxic materials, such as platinum and stainless steel, never of gold, silver, and copper alloys. (For various reasons, reverse stimulation through EMG electrodes is usually inadvisable.) The regulated variable should be current, not voltage, and no net direct current should be permitted. Multiple electrodes may be inserted and activated in desired sequence in alert and anaesthetized animals, but caution is required so that only the local muscles, rather than adjacent muscles and nerve trunks, be activated. Sometimes stimulation may be coupled with previous deinnervation. The effects of stimulation are usefully recorded on film or tape. The record only indicates the 'possible' actions of one or multiple muscles, but not the way they may be used by the animal.

The analyses of electrode reliability, sampling parameters, and correlation represent interactive stages of a single process. One attempts to establish which parts of which muscles are involved, what are their action patterns (the biological roles or functions in which they are involved), and to confirm that their participation is indeed necessary.

## 5. The final product

EMG analysis can provide a fairly simple set of graphs and values that permit one to discern which muscles are active when and how, while the organism performs its roles. A successful analysis should conclude by demonstrating the nature of the data and the logic of one's argument. It is useful to document that the approach was justified, that it was technically well-founded, and that the results were unequivocal. Such points are sometimes taken for granted and results are presented as if justified by the use of EMG itself.

The necessity of the approach demands at least the suggestion that the question has significance and that the results could not be gained by simpler means. Significance may demand only proof that no such process or organism has previously been studied. This rationale is a more powerful query than the consideration that a patently inappropriate hypothesis has here been 'falsified'. Demonstration of technical adequacy demands description of the kinds and numbers of organisms and replicates, of the nature of the equipment, and of the settings of filters and amplifiers, and of such arcane details as the signal-to-noise ratio. Inclusion of at least one record of the 'raw' (unretouched) EMG trace helps to convince properly sceptical colleagues.

Any results must incorporate adequate controls and at least should demonstrate good levels of correlation. They should include recordings of the major muscles shown to have been performing the actions. However, the simultaneous actions (or lack thereof) of possibly synergistic and antagonistic muscles also deserve demonstration. Some joints are bridged by multiple muscles in parallel or the systems proved otherwise complex. Then it remains necessary to show that the observed actions could have been performed only by the muscles for which this is claimed, and that these are sufficient to generate the actions. Finally, remember that performance of biological roles represents animal behaviour. Consequently, individual variability needs to be taken into account; EMG analysis must consider the behavioural history of the individual animal. Inter-individual variation, as well as that among populations (species, genera, families) needs consideration.

The secret of successful EMG recording lies less in the technical complexity of the apparatus than in an understanding of the signals, of the electrodes, and their bases. Physiological laboratories, even physiological teaching facilities, remain filled with preamplifiers, recorders, and tape systems that will be sufficient for simple experiments. Simple test, preliminary calibration, and minor modification are often sufficient for pilot experiments that may refine the areas for more detailed subsequent studies.

## Acknowledgements

The author thanks A. S. Gaunt, G. C. Gorniak, and G. E. Loeb for trenchant comments on the typescript, which was prepared with the assistance of a grant from the Leo Leeser Foundation.

## References

1. Brown, P. B., Franz, G. N., and Moraff, H. (1982). *Electronics for the modern scientist.* Elsevier, New York/Amsterdam.
2. Horowitz, P. and Hill, W. (1980). *The art of electronics.* Cambridge University Press, Cambridge.
3. Basmajian, J. V. and de Luca, C. J. (1985). *Muscles alive. Their functions revealed by electromyography* (5th edn). Williams & Wilkins, Baltimore/London.
4. Loeb, G. E. and Gans, C. (1986). *Electromyography for experimentalists.* University of Chicago Press, Chicago.
5. Stålberg, E. (1986). *Crit. Revs. Clin. Neurobiol.,* **2**(2), 125.
6. Hník, P., Vejsada, R., and Kassicki, S. (1988). *Rozpravy Ceskoslov. Akad. Véd. Matem. Prirod. Ved.,* **98**(2), 5.
7. Prosser, C. L. (1980). *Comparative animal physiology* (3rd edn). W. B. Saunders, Philadelphia.
8. Gans, C. (1982). *Exercise Sport Sci. Revs.,* **10**, 160.
9. Gans, C. and de Vree, F. (1987). *J. Morphol.,* **192**(1), 63–85 (April); erratum, **193**(3):323.
10. Gans, C. and de Gueldre, G. (1991). In *Physiological ecology of the amphibia* (ed. M. Feder and W. Burggren), University of Chicago Press, Chicago.
11. Guthe, K. (1981). In *Biology of the Reptilia,* Vol. 11 (ed. C. Gans and T. S. Parsons), p. 265. Academic Press, London.
12. Suzuki, A., Tsuchiya, T., Ohwada, S., and Tamate, H. (1985). *J. Morphol.,* **185**, 145.
13. Crowe, M. T. (1987). *Am. Zool.,* **27**(4), 1043.

14. Burke, R. E. (1977). In *The handbook of physiology: The nervous system II*, p. 345. US Dept. of Health and Human Services, Bethesda, Maryland.
15. Loeb, G. E., Pratt, C. A., Chanaud, C. M., and Richmond, F. J. R. (1987). *J. Morphol.*, **191**, 1.
16. Jayne, B. C., Lauder, S. V., Reilly, S. M., and Wainwright, P. C. (1990). *J. exp. Biol.*, **154**, 557.
17. Gans, C. (1974). *Biomechanics*. University of Michigan Press, Ann Arbor, NY.
18. Gordon, A. M., Huxley, A. F., and Julian, F. J. (1966). *J. Physiol.*, **184**, 170.

# 9

# Hydrostatic skeletons and muscular hydrostats

WILLIAM M. KIER

## 1. Introduction

### 1.1 Description of hydrostatic skeletal support

A hydrostatic skeleton is a fluid mechanism that provides a means by which contractile elements may be antagonized (1). Hydrostatic skeletons occur in a remarkable variety of organisms with examples not only from invertebrates but also from vertebrates. A hydrostatic skeleton is typically considered to include a liquid-filled cavity surrounded by a muscular wall reinforced with connective tissue fibres. This form of hydrostatic skeleton is seen, for example, in cnidarian polyps, holothuroid echinoderms, echinoderm water vascular systems, many molluscan bodies and organs and annelid, sipunculid, nemertean, and nematode worms (for reviews see refs. 1–7). Comparison of a variety of hydrostatic skeletal support systems shows that the extent and volume of the liquid-filled cavity is variable. In particular, recent work has identified a number of hydrostatic skeletons, termed muscular hydrostats, that consist of a tightly packed three-dimensional array for muscle fibres (8, 9). Examples of muscular hydrostats include the arms and tentacles, fins, suckers and mantles of cephalopod molluscs, a variety of molluscan structures, the tongues of many mammals and lizards, and the trunk of the elephant.

### 1.2 Arrangement of muscle fibres and movement in hydrostatic skeletons

The basic arrangement of muscle fibres in vermiform hydrostatic skeletons typically includes both circular and longitudinal muscle fibres. The hydrostatic skeleton provides a means by which these two muscle fibre orientations can antagonize one another, producing a variety of movements including elongation, shortening, and bending. The function of the hydrostatic skeletal support system relies on the fact that the enclosed liquid-filled cavity, typically a coelom, is constant in volume. Any decrease in one dimension must therefore result in an increase in another. Thus, to create elongation of the body, the circular muscles contract, decreasing the diameter and thereby increasing the length. Shortening of the body involves contraction of the longitudinal musculature. As the body shortens, its diameter must increase, with the resulting restoration of resting length in the circular muscles. The longitudinal and circular muscles are therefore antagonists.

In addition to the basic circular and longitudinal muscle arrangements, other muscle fibre arrangements are observed and allow additional types of movement. Four general categories of movement are possible: elongation, shortening, bending, and torsion. Elongation requires muscle fibres that are arranged such that their contraction decreases the diameter of the body or organ. In addition to the circular muscle fibres discussed above, two other muscle fibre arrangements create elongation: transverse muscle and radial muscle. Transverse muscle fibres extend across the diameter in parallel sheets and are seen, for example, in mammalian tongues, and the arms and tentacles of octopus and squid (8, 10). Radial muscle fibres radiate from the central axis in planes perpendicular to the long axis. Examples of radial muscle arrangements are observed in the elephant trunk and in the tentacles of the chambered nautilus (8, 11).

Shortening occurs as a result of contraction of longitudinal musculature and provides a means by which the muscle responsible for elongation may be antagonized. The relation between the diameter and length of a constant volume hydrostatic system allows amplification of the displacement or the force produced by the musculature responsible for elongation and shortening. This amplification is analogous to that produced in hardened skeletal elements in which joints and lever arms provide for leverage. For example, in a hydrostatic body or organ that is initially elongate—that is, one with a high length/diameter ratio—a relatively small decrease in diameter results in a large increase in length (see refs. 8, 10, and 12 for details). This means that the displacement and velocity generated by the musculature responsible for elongation is amplified. This amplification is significant, for example, in hydrostatic organs that are rapidly elongated such as the tentacles of squid, or in organs that are protruded over long distances such as the tongues of lizards and snakes (8, 9). Conversely, if the cylinder has a low length/diameter ratio, then a relatively small decrease in length creates a large increase in diameter. In this case force rather than displacement is amplified. This type of amplification may be of importance, for example, in anchoring individual segments of burrowing metameric worms. In the cases described above, the mechanical advantage of the antagonistic musculature is opposite; that is, if the hydrostatic skeletal system provides for amplification of *displacement* by a group of muscles, then the *force* of the antagonists will be amplified.

Bending movements require contraction of longitudinal muscle along one side of the body or organ. The bending moment is greatest if the longitudinal muscle is peripherally arranged—that is, located as far from the central axis as possible. In order for this longitudinal muscle contraction to create bending, some component of the body or organ must resist the longitudinal compressional force that would otherwise cause shortening. In some hydrostatic skeletons, connective tissue fibres provide resistance to longitudinal compression (see below) while in others, muscle provides the resistance. Since any decrease in length due to a longitudinal compressional force must result in an increase in diameter of a hydrostatic system, muscles arranged to control diameter (circular, transverse, and radial) can provide the resistance to longitudinal compression required for bending (8). Note that these are the same muscle arrangements that produce elongation. In the case of bending, however, these muscles operate synergistically with the longitudinal muscles rather than antagonistically.

The final category of movement in hydrostatic skeletons is torsion or twisting around the long axis. The muscles responsible for this movement are arranged in helical layers around the body or organ. For torsion in either direction to be possible, both right- and left-handed helical muscle layers must be present. The torsional moment of these muscle layers is maximized if the layers are located toward the outer surface, as far from the central axis as possible. Indeed, the helical musculature that has been observed in hydrostatic skeletons typically wraps the remainder of the more central musculature. The fibre angle (the angle that a helical fibre makes with the long axis of the structure) of the helical muscles affects the forces that are exerted on the hydrostatic system as the helical contracts. If the helical muscles are arranged at a fibre angle of 54°44′, then their contraction will generate a torsional force without affecting the length or diameter of the cylinder. If, however, the fibre angle of the helical muscles is greater than 54°44′, then their contraction will create both a torsional force and a force that will tend to decrease the diameter and thereby increase the length of the cylinder. If the fibre angle is less than 54°44′, contraction of helical muscles will create both a torsional force and one that tends to shorten the cylinder. Helically-arranged muscle has been observed in the arms and tentacles of cephalopods, in some lizard tongues, and in the elephant trunk (8, 10, 11).

## 1.3 The role of connective tissue fibres in movement and changes in shape

Connective tissue fibres often play a crucial role in the determination of the range and type of movements possible in hydrostatic skeletons. The most prevalent arrangement of connective tissue fibres in hydrostatic skeletons is that of the crossed-fibre helical fibre array in which sheets of connective tissue fibres are arranged in both right- and left-handed helixes, wrapping the hydrostatic body or organ. The mechanical implications of these relatively inextensible fibres have been analysed for a number of vermiform hydrostatic skeletons. The fibre angle of the connective tissue fibres plays a crucial role in determining the range of shape change that is possible. This has been explored with a geometrical model of a cylinder wrapped with a single constant length helical fibre. The model compares the enclosed volume as a function of the fibre angle (angle that the fibre makes with the long axis of the cylinder). As the fibre angle approaches 0° and 90° the volume of the cylinder approaches zero. A maximum volume occurs between these two extremes at a fibre angle of 55°44′. This model and various elaborations of it have been applied to the analysis of a number of hydrostatic skeletal support systems including, for example, nemertean, turbellarian, and nematode worms and the tube feet of echinoderms (13–17). These studies have demonstrated the importance of the crossed fibre array in controlling the range of shape possible in these systems.

In addition to crossed fibre helical arrays, other arrangements of connective tissue fibres have also been shown to serve important roles in hydrostatic skeletal support systems. In the mantle and fins of squid, connective tissue fibres are observed to be embedded in the tightly packed musculature of these structures, with the connective tissue fibres at an angle to the muscle fibres. These intermuscular connective tissue fibres control changes of shape, provide for muscular antagonism, and may serve in elastic energy storage, increasing the efficiency of movement (18–23).

## 1.4 Analysis of hydrostatic skeletal support

In hydrostatic support systems, the arrangement of the muscle and the connective tissue fibres determines the types of movement, range of movement, and changes of shape that are possible. Features such as connective tissue fibre orientation and three-dimensional course of muscle bundles have significant impact on mechanics and function. Thus, analysis of hydrostatic skeletal support systems requires a detailed analysis of the morphology of the musculature and connective tissues at a microscopical level. Much of this chapter will be concerned with the techniques used in morphological analysis at this level. Further analysis of hydrostatic skeletons requires direct documentation of the sequence and duration of muscle activity. The most convenient method is that of electromyography, reviewed in this volume by Gans (Chapter 8). Special electrode techniques are often required for electromyography of hydrostatic skeletons and are described below. The force produced by contractile activity of the musculature is transmitted as changes in pressure within hydrostatic skeletons and thus, techniques for measuring pressure of hydrostatic skeletons are important and are described below. Analysis of movement is also required and Chapter 3 in this volume, by Biewener and Full, provides details of kinematic analysis. A convenient method for kinematic analysis of aquatic animals is also described below; it is of particular use in the study of hydrostatic skeletons.

# 2. Morphological analysis

## 2.1 Introduction

Because the form of many hydrostatic skeletal systems is relatively poorly studied, one of the first steps in their analysis is a complete and detailed morphological study. This analysis involves a mixture of classical and recent morphological techniques. Although a technique such as paraffin histology might be assumed to have little relevance in today's world of DNA technology, it actually provides critical data concerning the microanatomy of hydrostatic skeletal support systems. Furthermore, today's students are in fact more likely to be familiar with modern molecular biological techniques than with morphological techniques. Therefore considerable attention will be devoted to the relative advantages and disadvantages of various morphological techniques, and to details concerning their use. The focus of this discussion will be on techniques useful for the analysis of structure at the light microscopic level. Electron microscopical analysis of hydrostatic skeletons is a more specialized application and a detailed description of the technique is beyond the scope of this chapter. Excellent descriptions of electron microscopical techniques are available (24).

The most useful techniques for morphological analysis of hydrostatic skeletons at the light microscopic level are sections of tissue embedded in paraffin or in glycol methacrylate (GMA) plastic. The choice of technique depends on both the specimen and the level of analysis. Tissue embedded in GMA plastic is subject to greatly reduced distortion and artefact (relative to paraffin) and can be sectioned at 0.5–3.0 μm, providing morphological detail that may rival low power electronmicroscopical analysis. Nevertheless, paraffin histology still serves an important role in morphological analysis. Most hydrostatic skeletal support systems are relatively large. In order to accurately evaluate the morphological components, it is typically necessary to section serially the structure in three mutually perpendicular planes. This is often the only way to obtain a grazing section and therefore observe the critically important connective tissue fibre components of the structure. Because paraffin sections are typically made at 5–10 μm, fewer sections are required in order to section through an entire structure, and thus there are fewer sections to mount and stain and fewer slides to mount coverslips on. For considerations of general microanatomy and tissue component arrangement, the additional resolution provided by plastic is unnecessary. Furthermore, paraffin forms serial ribbons with ease. Ribbons greatly facilitate the production of serial sections; successive sections are aligned relative to one another, making it a simple matter to affix them in the proper order on microscope slides. The resulting linear alignment also aids in observing the sequence of sections on the microscope. Although there are techniques that allow the production of serial ribbons with glycol methacrylate, in my experience it is much more difficult to obtain ribbons with plastic. Assembling individual sections in a complete series on slides is a difficult, time-consuming, and painstaking process.

In addition to sectioning considerations, paraffin offers some distinct advantages in terms of staining of tissue components. A vast array of staining techniques are available for paraffin sectioned material. Those provided below are several of the ones that the author has found to be particularly useful. In particular, brilliant and beautiful staining techniques are available that show dramatic differentiation between various tissue components, especially between connective and muscle tissue. These stains are invaluable for an initial morphological analysis. Some of the techniques originally developed for paraffin can be adapted for staining of glycol methacrylate plastic sections but the results are often unpredictable.

After the microanatomy and arrangement of tissue components of the entire structure or body has been examined with paraffin sections, additional morphological detail in specific areas can be provided by GMA sections. Distortion and shrinkage is minimal in GMA embedded material, in part because heat is not required during processing of the tissue. In addition, the thinner sections obtained with the harder plastic embedding material allow much higher resolution of specimen detail (25, 26). Low-melting-point polyester wax (27) can also be used to reduce distortion and shrinkage due to heat but it is often difficult to adhere polyester wax sections to slides. Coating the sections, once on the slide, by dipping in a 0.1% solution of pyroxylin in 50% ether/50% ethyl alcohol helps to prevent them from floating off in the solutions but increases staining times considerably. In general, it is best to perform the initial morphological study with serial paraffin sections and then to use GMA sections for additional resolution of morphological details. This approach allows one to exploit the advantages of each technique.

Frozen sections of unfixed material are sometimes useful in providing an initial morphological survey of connective tissues, particularly if being viewed with polarized light microscopy. Nevertheless, a complete analysis with serial sections is usually required in order to analyse all tissue components and it is therefore more convenient to bypass frozen sections and begin with serial paraffin sections.

In the procedures that follow, several general considerations should be borne in mind. Most of the steps in processing tissue for plastic or paraffin embedding rely on diffusion. Difficulties experienced in these methods can usually be traced to lack of sufficient infiltration of the tissue by one component or another. Thus, one should always attempt to minimize the size and maximize the surface to volume ratio of the specimen. Often, it is possible to trim a specimen into a 'slab' rather than a cube. This strategy provides a full-size block face but greatly reduces diffusion distances. In addition, it is useful to augment diffusion with convection, typically by placing the vial of solution with tissue in a tissue rotator. In situations where rotating the specimen is impractical, one should consider the specific gravities of the component to be removed from the tissue compared to that of the bathing solution and suspend the tissue accordingly. For instance, if the component to be removed is more dense than the bathing solution, the exchange will be augmented if the tissue is suspended in the bathing solution rather than resting on the bottom of the vial.

## 2.2 Fixation

The choice of fixative depends on the intended embedding procedure. With the advent of preparative techniques for electron microscopy, a variety of fixative procedures have been developed that provide remarkable preservation and minimize distortion, shrinkage, and artefact. These procedures, however, are generally applied to small blocks of tissue and thus, penetration of the tissue by the fixative is less of an issue. For the relatively large blocks of tissue that must be fixed for the examination of the morphology and micro-anatomy of hydrostatic skeletons by paraffin embedding, the more classical and simple fixatives seem to be superior because many are excellent in terms of penetration of the tissue. In addition, many of the fixatives developed for electron microscopy include glutaraldehyde as a component and glutaraldehyde often interferes with the staining procedures used for paraffin embedded tissue. For paraffin embedding, considerations of fixative osmolality seem to be less important, because the distortions and shrinkage induced as a result of the heat during the embedding procedures are more significant than distortions due to osmolality differences. For embedding in glycol methacrylate, the fixatives developed for electron microscopy are superior. These give excellent preservation with little distortion and since the tissue blocks are generally smaller, their reduced penetration is less of a problem. Adjustments to fixative osmolality may improve preservation. Below are several fixative recipes for paraffin and glycol methacrylate embedding.

Since the goal of fixation is to preserve the tissue in a state that is as close as possible to that of living tissue, care must be taken to minimize post-mortem changes. It is thus necessary that the tissue be fixed as soon as possible after death, preferably in chilled fixative. Perfusion of the tissue with fixative is a help in many cases. In most cases narcotization is required. For marine animals, a 1:1 mixture of sea water and 7.5% $MgCl_2 \cdot 6H_2O$ is excellent. Other useful narcotization methods can be found in Pantin (28) and Humason (29). The comments above concerning diffusion and surface to volume ratio are of particular importance during fixation. As a general guide-line, a fixative volume to tissue volume ratio of 10:1 or greater is recommended.

A comment concerning formalin is necessary at this point. The term 'formalin' refers to a saturated solution of formaldehyde in water, typically 36–38% formaldehyde. Fixation protocols refer typically to concentrations of formalin rather than formaldehyde. When preparing fixatives, one simply considers formalin to be 100% and then makes the appropriate dilutions. If formalin is stored for extended periods, polymerization and oxidation of formaldehyde occurs. The presence of a white precipitate suggests that the formalin will be less effective.

### 2.2.1 Buffered formalin

For initial morphological analysis, this simple fixative is an excellent choice. It works well with paraffin embedding and staining procedures. Substitute sea water for distilled water for marine specimens. If fixing marine animals in the field where simplicity of procedure is desirable, excellent results can be obtained with 10% formalin in sea water without the sodium phosphate.

The action of formalin is progressive, that is, fixation improves over time and the tissue can be left in fixative for extended periods. Fix for a minimum of 48 hours.

- formalin                                           100 ml
- distilled water (sea water, if marine)             900 ml
- sodium phosphate, monobasic ($NaH_2PO_4 \cdot H_2O$)    4.0 g
- sodium phosphate, dibasic ($Na_2HPO_4$)            6.5 g

### 2.2.2 Alcoholic Bouin's (28)

For tissues difficult to penetrate, for example dense muscular tissue, Bouin's fixative often works well. The shrinkage that occurs during fixation can be considerable and thus, it is useful only for paraffin embedded material. It cannot be used for organisms where preservation of calcareous inclusions is required because they are dissolved by the acid. Fix for 24 hours but remove soon after because tissue left in Bouin's tends to become over hardened. After fixation, transfer to 95% ethyl alcohol. Because picric acid can interfere with staining, the alcohol should be changed until most of the excess picric acid is removed.

- picric acid            1.0 g
- acetic acid            15 ml
- formalin               60 ml
- 80% ethyl alcohol      150 ml

### 2.2.3 Phosphate-buffered glutaraldehyde (30)

This is an excellent general fixative for use with GMA embedding procedures. Although the preservation is excellent and little distortion occurs, the fixative penetrates poorly. Tissue blocks should not be thicker than 1–2 mm. It is useful to maintain a 1.0 M solution of sodium chloride if you wish to modify the osmolality of the fixative. The 1.0 M solution can be diluted to the appropriate strength and used in place of the 0.34 M solution in the formula below. Fix for 2–4 h. After fixation, rinse 15–30 min in a 1:1 solution of 0.4 M phosphate buffer and 0.6 M (7.01 g in 200 ml water) sodium chloride.

- 25% glutaraldehyde                                          5 ml
- 0.34 M sodium chloride (3.97 g in 100 ml)                   20 ml
- 0.4 M Millonig's phosphate buffer (see below)              25 ml

Millonig's phosphate buffer (0.4 M):

- sodium phosphate (monobasic)                               11.08 g
- sodium hydroxide                                           2.85 g
- distilled water to make                                    200.0 ml

## 2.3 Decalcification

Tissue with mineralized inclusions must be decalcified prior to dehydration and clearing. Although a variety of recipes for decalcifying solutions are available in histology texts, the decalcifying solutions sold commercially that include chelating agents are excellent and convenient. It is essential that the tissue be washed in several changes of water following decalcification. Following the wash, one can then proceed directly to the dehydration series.

## 2.4 Dehydration and clearing

Since water and paraffin are not miscible, the tissue must be completely dehydrated before it can be embedded. Additionally, since dehydration is most easily performed with ethyl alcohol, which is also not miscible with paraffin, an intermediate 'clearing' agent must be used that is miscible in both alcohol and paraffin. Poor histological results can most commonly be traced to a failure to completely dehydrate and then remove alcohol from the tissue. In the past, xylene was employed commonly as a clearing agent. In response to concerns about the health hazards associated with the use of xylene, non-toxic clearing agents have been developed. These clearing agents (for example, Histoclear, National Diagnostics, Somerville, NJ, USA; Hemo-De, Fisher Scientific Products, Pittsburg, Pennsylvania, USA) can be substituted for xylene with no change in procedure, seem to cause less hardening of the tissue, and provide excellent staining. Their use is highly recommended.

The amount of time required in each step of the dehydration and clearing process varies from tissue to tissue and must therefore be determined by experimentation. For tissue blocks 3–5 mm on a side, the author uses 1-hour steps of the following series as a starting point (all steps on tissue rotator): 30% ETOH, 50% ETOH, 70% ETOH, 95% ETOH, 100% ETOH (3 changes), 50/50 100% ETOH/clearing agent, clearing agent (2 changes). Tissue fixed in alcoholic Bouin's will be in 95% ETOH and can be started in the series at the 100% ETOH step. After the second clearing agent bath, transfer the tissue to a vial filled with paraffin chips and clearing agent and allow the tissue to remain overnight before beginning the infiltration series in the oven the following day. These times can be shortened considerably in some instances.

Dehydration for embedding in glycol methacrylate plastic is usually performed in a graded ethyl alcohol series. Complete dehydration is unnecessary, however, since glycol methacrylate is miscible with water. No clearing agent is required. Shorter times are required than those listed for paraffin above, because the tissue blocks are typically smaller. As a starting point, 30-min steps of the following series are recommended: 30% ETOH, 50% ETOH, 70% ETOH, 95% ETOH. After the 95% ETOH step, the tissue can be transferred directly to unpolymerized GMA. See below. One can also eliminate the alcohol series and transfer the tissue directly to several changes of glycol methacrylate. This use of glycol methacrylate as the dehydrating agent is more expensive than dehydration in alcohol but yields superior results in some instances.

## 2.5 Infiltration and embedding

Successful embedding requires complete infiltration of the tissue with molten paraffin or unpolymerized GMA. In the case of paraffin embedding, complete removal of the clearing agent is necessary. This is accomplished by transferring the tissue through several baths of molten paraffin. As a starting point, three paraffin infiltration baths, 1 hour each should be tried. It is convenient to keep the paraffin baths arranged and labelled sequentially in the oven in order to ensure that the final infiltration bath contains as little clearing agent as possible. A vacuum oven is useful because reduced pressure is often required for complete infiltration of the tissue with paraffin.

The commercially available paraffin embedding media (for example, Paraplast Plus, Monoject Scientific, St Louis, Missouri, USA; Ameraffin, Baxter Healthcare, American Scientific Products Division, McGraw, Illinois, USA) work well and include additives that are claimed to aid sectioning and infiltration. Care should be taken in setting the paraffin oven because temperatures more than a few degrees above the melting point are said to reduce the effectiveness of the additives.

Due to the use of paraffin embedding in pathology, a variety of embedding aids and supplies are available. The most useful are disposable plastic embedding boats and embedding cassettes and moulds. The disposable embedding boats are available in a variety of sizes, can be labelled with indelible markers and reduce time in trimming the solidified blocks. The embedding cassettes and moulds have all of the advantages of the disposable boats and, in addition, eliminate the step of mounting the block because the cassette can be clamped into the microtome directly. Mould release compounds are available that help to free the solidified block from the mould.

Once the tissue has been placed in the molten paraffin in the mould and oriented properly, heated forceps are used to release all bubbles from the tissue. The paraffin should then be solidified quickly by placing the mould in a water bath. Blocks that solidify slowly in air do not section as well.

Successful GMA embedding requires complete infiltration of the tissue followed by complete polymerization of the GMA monomer. Three infiltration steps of 30 min duration on a tissue rotator are recommended. Two aspects of the polymerization are of particular importance: heat and oxygen. For most GMA formulations, oxygen inhibits the polymerization and must therefore be excluded during polymerization. In addition, the polymerization is exothermic and, in larger blocks, the excess heat produced can create distortions in the tissue similar to those produced by the heat of paraffin embedding. Butler (31) describes a simple chamber that can be purged of oxygen during polymerization and is easily constructed. The chamber incorporates an aluminium plate and water bath that serves to conduct heat away from the blocks during polymerization. As an alternative, Bennett *et al.* (25) recommend polymerization in an oil bath. In addition, embedding moulds and stubs that include heat sinks are available for GMA embedding. Recently, the author has obtained excellent results using a GMA formulation (Historesin, Reichert-Jung, Cambridge Instruments, Deerfield, IL, USA) that polymerizes in the presence of oxygen.

## 2.6 Sectioning

### 2.6.1 Tissue embedded in paraffin

With tissue that has been fixed and embedded carefully, serial paraffin sections of 5–10 μm are readily obtained. Humason (29) includes an excellent trouble-shooting chart for paraffin sectioning. If the tissue is embedded in a disposable mould, the mould is peeled away from the solidified block and the block is then mounted with molten paraffin on the appropriate microtome tab. With cassettes, the block is simply popped out of the mould. The block can then be trimmed with a razor blade. It is critical that the upper and lower sides of the block (that is, those parallel to the knife edge) be trimmed as close to parallel to one another as possible. If they are not trimmed parallel, ribbons of serial sections may be more difficult to obtain and the ribbons of sections will curve, making it difficult to mount as many sections on a slide and less convenient to study the series on the microscope. It is helpful to trim the block as a pyramid with the block face (plane of section) as a trapezoid (*Figure 1*). The trapezoidal shape aids in sectioning and provides a reference for the sequence of sections in a ribbon. The pyramid is useful because it allows retrimming of the block as sectioning proceeds, which aids in correcting non-parallel or damaged block sides.

The choice of knife for paraffin sectioning is somewhat subjective. A well-sharpened steel microtome knife will cut excellent sections, but honing and stroping a knife by hand is an art. Automated knife sharpeners work quite well and the disposable abrasive sheets (Thomas Scientific, Swedesboro, NJ,

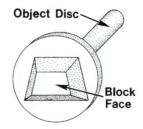

**Figure 1.** Diagram illustrating convenient method for trimming paraffin blocks. A block mounted on an object disc is shown. The object disc is clamped in the jaws of the microtome during sectioning. The block is trimmed as a pyramid with the block face (the plane of section) shaped as a trapezoid.

USA) that are now available reduce the mess and inconvenience of liquid honing compounds. A convenient alternative is a blade holder with disposable blades. Many of the disposable blades sold for the holders cut excellent sections and one simply replaces the blade when dulled or nicked. Glass Ralph knives (see below) can also be used for paraffin sectioning, although one must experiment with the profile of the edge to obtain good results.

Once a paraffin block has been sectioned, the ribbons of sections must be adhered to microscope slides. Care must be taken in cleaning the slides, otherwise the sections may float off during the staining procedures. Soaking the slides in acid alcohol (1% hydrochloric acid in 70% ethyl alcohol) does an excellent job of cleaning the glass surface. The clean, dry slide can then be coated with a thin layer of Mayer's albumen which is most conveniently obtained from commercial sources (Fisher Scientific Products, Pittsburg, Pennsylvania, USA). Paradoxically, the less albumen applied to the slide, the better the adherence of sections to the slides. Adhesion can also be provided by using 'subbed' slides (29), although the author has found albumen to be more suitable, and in preparing serial sections finds it most convenient to place water on the surface of the albumen-coated slide with an eye dropper and then to place the ribbon of sections on the drop of water. The slide can then be transferred to a hot-plate set at 40–50°C. The heat causes the sections to expand, removing much of the compression created during sectioning. Once the sections have expanded, excess water can be drained off the slides, and the sections can be precisely positioned with a brush or forceps. The slides are then allowed to dry on the hot-plate for approximately an hour. The heat is necessary for adhering the sections to the slides. The slides should be labelled clearly. Slides frosted on one end have been found to be most useful because they are easily labelled with pencil or India ink.

### 2.6.2 Tissue embedded in GMA

Sectioning of tissue embedded in GMA plastic requires a retracting microtome; that is, a microtome with a mechanism that moves the block to the side or away from the knife on the return stroke. Several manufacturers sell retracting microtomes designed specifically for GMA sectioning. If a project will involve extensive GMA sectioning, purchase of one of the GMA microtomes is recommended. Alternatively, small GMA blocks can be sectioned on an ultramicrotome. This is often an inexpensive solution because ultramicrotomes that have reached the end of their useful life for ultramicrotomy will nevertheless do an excellent job of cutting the much thicker GMA sections.

Glass knives work well for sectioning GMA blocks. For block faces smaller than 6 mm, triangular glass knives (31) are excellent. In addition, these knives are usually readily produced because most electron microscopy facilities include mechanized knife breakers that automate the production of triangular glass knives. These knife breakers allow one to produce quickly glass knives that are of excellent quality for GMA sectioning. The mechanized knife breakers can also be used with thicker glass strips for use with block faces as wide as 9 mm. For even larger block faces, the best results are obtained using Ralph knives. Ralph knives can be broken by hand as described in Bennett *et al.* (25). Alternatively, a number of mechanized Ralph knife breakers are now available that provide for more convenient and consistent production of Ralph knives. The author has had little success in adapting steel knives or disposable steel blades to the sectioning of GMA blocks. Nevertheless, steel knives designed specifically for GMA are now available and may be worth investigating.

Sectioning of GMA is usually done with a dry knife. It is helpful to lift and hold the bottom edge of the section and to maintain slight tension on the section as it comes off of the knife edge. This procedure helps to reduce folding of the sections. The sections are then transferred to the surface of a water bath where they flatten. Folds in the sections can usually be avoided by dropping the section on the water surface carefully. Once the section is floating on the surface, folds can sometimes be removed by probing with an eyelash or brush. Flattened sections are picked up on cleaned slides by inserting the slide at an angle through the water surface underneath the section and lifting the slide. Before removing the slide from the water, the position of the sections can be adjusted with a brush. The slide is then dried on a hot plate at 60 °C. The sections adhere to clean slides without application of any adhesive (25, 26).

## 2.7 Staining

A tremendous variety of staining procedures are available that allow identification of various tissue and cellular components in paraffin embedded tissue (28, 29). For the study of hydrostatic skeletons and muscular hydrostats, howver, the most useful are those that clearly differentiate muscle and connective tissue. The routine stains used by pathologists (for example, Hematoxylin and Eosin), are of little use in this regard. K. Smith (Duke University) and I have tried a variety of stains and have found several excellent protocols that give beautiful results and provide clear differentiation of muscle and connective tissue. These protocols are highly recommended and are listed below.

Staining of paraffin sections typically involves aqueous stains. The paraffin must therefore be removed from the sections using the clearing agent (2 changes, 5 min each), the clearing agent must be removed with 100% ethyl alcohol (2 changes, 5 min each) and then finally, the alcohol must be removed and the sections rehydrated (95% ETOH, 70% ETOH, 50% ETOH, 30% ETOH, distilled water: 5 min each). Since examination of hydrostatic skeletons and muscular hydrostats requires serial sections, these steps and the staining process itself, involve many slides. It is thus most convenient to use 8 × 10 × 8 cm glass staining trays with glass carriers that hold multiple slides. A number of slides can be processed quickly and variation in staining between slides is reduced.

Bennett *et al.* (25) provide protocols and information on staining of sections embedded in glycol methacrylate plastic. The most useful and convenient is Lee's methylene blue–basic fuchsin stain for GMA sections (*Protocol 1*). The thinner sections possible with GMA embedding appear to be much less intensely stained than thicker sections. One must therefore become accustomed to studying sections that show weaker staining effects (25).

## 2.8 Staining protocols

### 2.8.1 Milligan trichrome for paraffin sections, adapted from ref. 29

This is an excellent stain for hydrostatic skeletons because it clearly differentiates muscle (magenta) and collagen fibres (blue with aniline blue and green with fast green). Nuclei stain magenta and red blood cells stain orange to orange red.

---

**Protocol 1.** Preparation of stock solutions

**1.** Mordant
   (Mix 3 parts of solution A with one part solution B below; use within 4 h)
   Solution A, Mix:
   - potassium dichromate (carcinogen-mix in hood)   3.0 g
   - distilled water                                 100.0 ml

   Solution B, mix:
   - hydrochloric acid, concentrated                 10.0 ml
   - 95% ethyl alcohol                               100.0 ml

2. Acid fuchsin, mix:
   - acid fuchsin (C.I. 42685)                        0.1 g
   - distilled water                                  100.0 ml

3. Phosphomolybdic acid solution (1%), mix:
   - phosphomolybdic acid                             2.0 g
   - distilled water                                  200.0 ml

4. Orange G, mix:
   - orange G (C.I. 16230)                            2.0 g
   - 1% phosphomolybdic acid                          100.0 ml

5. Aniline blue (fast green may be substituted for aniline blue)
   Stock solution, mix:
   - aniline blue (C.I.42755)                         10.0 g
   - 2% acetic acid (2 ml/98 ml distilled water)      100.0 ml
   Working solution, mix:
   - aniline blue stock                               10.0 ml
   - distilled water                                  90.0 ml

---

**Protocol 2.** Staining procedure for Milligan trichrome stain

The following procedure includes staining times that the author has found to work well, but experimentation may be necessary depending on the tissue, fixative, etc. Consult Humason (29) for additional information.

1. Deparaffinize and transfer slides through 100% ethyl alcohol into 95% ethyl alcohol.

2. Mordant slides in potassium dichromate–hydrochloric acid solution for 5 min.

3. Rinse in distilled water.

4. Stain in acid fuchsin: 30 sec.

5. Rinse in distilled water.

6. Fix stain in phosphomolybdic acid solution: 2 min.

7. Stain in orange G: 30 sec.

8. Rinse in distilled water.

9. Treat with 1% aqueous acetic acid (1 ml/99 ml distilled water): 2 min.

10. Stain in aniline blue: 45–60 sec.

11. Treat with 1% acetic acid: 3 min.

12. Rinse in 70% ethyl alcohol. Transfer to 95% ethyl alcohol: 5 min.

13. Finish dehydration in absolute alcohol, 2 changes: 3 min each.

14. Clear and mount coverslip.

### 2.8.2 Picro-ponceau with haematoxylin for paraffin sections (29)

This is also an excellent, beautiful stain that clearly differentiates between muscle (yellow) and connective tissue fibres (red). Nuclear and nervous tissue differentiation is much greater with this stain than with Milligan's, as both stain dark brown to black. Elastic fibres, red blood cells, and epithelia stain yellow.

---

**Protocol 3.** Preparation of stock solutions for picro-ponceau with haematoxylin

1. Picro-ponceau stain, mix:
   - ponceau S (C.I. 27195), 1% aqueous          10.0 ml
   - picric acid, saturated aqueous              86.0 ml
   - acetic acid, 1% aqueous                      4.0 ml
2. Weigert iron hematoxylin
   (Mix solutions A and B below. The solution will turn black, and can be used for 1–2 weeks.)

   Solution A, mix:
   - ferric chloride ($FeCl_3 \cdot 6H_2O$)      2.5 g
   - ferrous sulfate ($FeSO_4 \cdot 7H_2O$)      4.5 g
   - hydrochloric acid, concentrated             2.0 ml
   - distilled water                           298.0 ml

   Solution B, mix:
   - haematoxylin                                1.0 g
   - 95% ethyl alcohol                         100.0 ml

---

**Protocol 4.** Staining procedure for picro-ponceau and haematoxylin

1. Deparaffinize, transfer through alcohol and hydrate slides down to water.
2. Overstain in haematoxylin: 5–10 min.
3. Wash in running water until slides are deep blue: 5–10 min.
4. Stain in picro-ponceau: 3–5 min. Picro-ponceau acts as both a stain and as a destaining agent on haematoxylin. Its action should be monitored by rinsing the slides for a few seconds in distilled water and checking under the microscope. The nuclei should be sharp.
5. Dip several times in 70% ethyl alcohol.
6. Dehydrate in 95% ethyl alcohol, 2 changes, to remove excess picric acid.
7. Dehydrate in 100% ethyl alcohol, clear, and mount coverslip.

---

**Protocol 5.** Lee's methylene blue–basic fuchsin stain for GMA
sections

1. Prepare the following stock solutions:
   (a) $4 \times 10^{-3}$ M methylene blue
   - methylene blue (C.I. Basic Blue 9, C.I. No. 52015)   0.13 g
   - deionized water                                       100 ml
   (b) $4 \times 10^{-3}$ M basic fuchsin
   - basic fuchsin (C.I. Basic Red 9, C.I. No. 42500)      0.13 g
   - deionized water                                       100 ml
   (c) 0.2 M phosphate buffer (pH 7.2–8)
   (Use protocol for Millonig's phosphate buffer; see section 2.2.3,
   phosphate buffered glutaraldehyde.)

2. Prepare staining solution by mixing:
   - $4 \times 10^{-3}$ M methylene blue    12 ml
   - $4 \times 10^{-3}$ M basic fuchsin     12 ml
   - 0.2 M phosphate buffer                 21 ml
   - 95% ethyl alcohol                      15 ml
   - Filter after mixing. Useful for 4–5 days.

3. Stain for 10–15 sec, then dip slide in distilled water briefly, dry with a blast
   of inert gas, and mount coverslip.

---

# 3. Microscopy

## 3.1 Brightfield microscopy

Simple brightfield microscopy is most useful for the initial examination of
both paraffin and plastic sections. Oculars with high field number (for example,
23–25) are particularly convenient because they provide a wide field of view
and thus allow a more rapid survey of the sections. Contrast enhancing
methods such as phase-contrast and differential inference contrast may be
helpful in some instances.

## 3.2 Polarized light microscopy

Polarized light microscopy is of particular use in the examination of hydrostatic skeletons. One of the most important aspects in the analysis of hydrostatic skeletons concerns the arrangement and orientation of connective tissue
fibres. The connective tissue fibers are generally collagen. Because of the
alignment of the collagen molecules, the fibres are birefringent and are
therefore amenable to analysis with polarized light. Indeed, in many instances,
it is extremely difficult to visualize the crossed fibre arrays of hydrostatic
skeletons without polarized light microscopy.

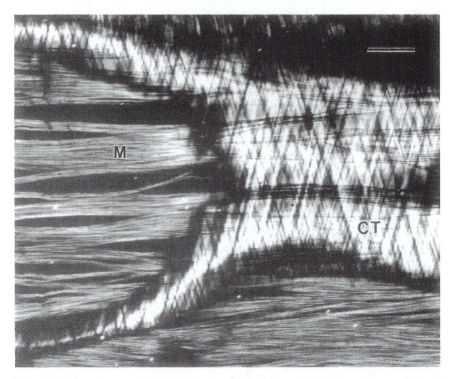

**Figure 2.** Photomicrograph of 7 μm frontal paraffin section of the arm of the squid, *Sepioteuthis sepioidea*, viewed with polarized light microscopy. With polarized light, the arrangement of the birefringent connective tissue fibres (CT) is apparent. The muscle fibres in the plane of the section (M) are also birefringent. For additional detail concerning the morphology, see ref. 10. The scale bar length equals 100 μm.

Polarized light microscopy is useful in the analysis of fibre arrays primarily in terms of visualizing the fibres, documenting their arrangement and measuring fibre angles (*Figure 2*). Birefringence can sometimes be used to distinguish between collagen fibres and rubber-like protein 'elastic' fibres, which are isotropic (19, 33). Crystallographic techniques such as measurement of the sign of the birefringence or obtaining crystal interference figures are unnecessary. The equipment required is therefore much less elaborate and expensive. While a polarizing microscope with a rotating stage, strain-free objectives, and high-quality polarizer and analyser is ideal, a good quality brightfield microscope can be adapted for simple polarized light microscopy. A polarizing filter, or even a piece of polarizing film, can be placed in a holder in the microscope condenser. A second polarizing filter serves as the analyser and can be inserted above the microscope nosepiece. The polarizing filter is then rotated in the condenser slot until the field is as dark as possible. If the stage is not rotatable, the slide holder can be removed and the slide itself can then be rotated on top of the stage. A first-order red filter is sometimes helpful in visualizing fibre arrangements but is not absolutely necessary.

Polarized light microscopy may be used on specimens stained according to the procedures outlined above, but it is often useful to deparaffinize several slides from the series, mount coverslips without staining and use the unstained section for polarized light analysis. For glycol methacrylate sections, a coverslip is simply mounted on the unstained section.

# 4. Analysis of sections

Once serial sections through an object have been obtained, the sections must be interpreted with respect to the three-dimensional arrangement of the structure. As discussed by Elias (34), misinterpretation of three-dimensional structures from the observation of sections is common in the literature. The recommendation above of obtaining serial sections in three mutually perpendicular planes is of particular importance in this regard. The perpendicular section planes are a tremendous aid in interpreting the three-dimensional arrangement of the structure because they allow one to check predictions based on observation in one plane with the observations from another plane. For complex structures, this process of checking predictions of a particular interpretation in other section planes requires considerable patience and time.

## 4.1 Computer-assisted three-dimensional reconstruction

Computer-assisted three-dimensional reconstruction of serial sections is often useful in the interpretation of complex morphology. Several three-dimensional reconstruction software packages are now available for microcomputers and the equipment required is therefore relatively inexpensive. The system used in my laboratory (PC3D, Jandel Scientific, Corte Madera, California, USA) has proved to be extremely useful, in particular for students learning morphological techniques. Individual sections are traced using a microscope equipped with a camera lucida and the tracings are then digitized. The software allows the different objects in the sections to be categorized and takes into account the magnification and distance between sections. Once all of the sections in a series have been digitized, the software then stacks the sections in three-dimensional space and displays a reconstruction that can be rotated in any direction. The reconstruction can include or exclude any of the categorized objects, they can be displayed in a variety of colours, and they can be reconstructed with or without hidden line elimination. See *Figure 3* and Kier and Smith (35) for examples. In addition, it is possible to plot two images of the object that differ in rotation and view the plots as a stereo pair. The resulting three-dimensional image is often extremely useful in resolving questions about the precise shape, arrangement, and location of objects in the

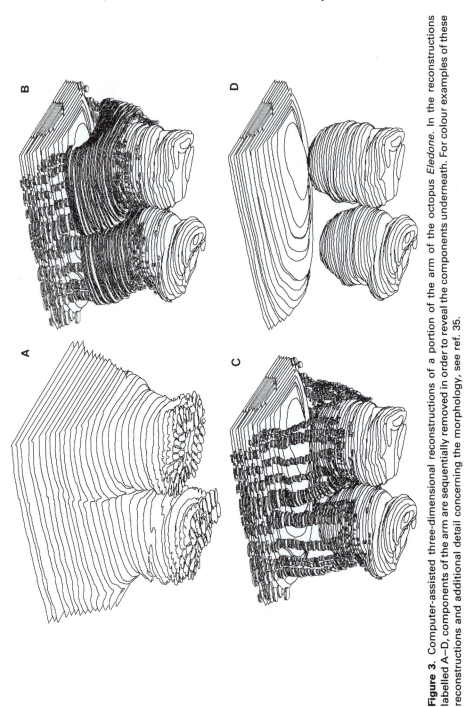

**Figure 3.** Computer-assisted three-dimensional reconstructions of a portion of the arm of the octopus *Eledone*. In the reconstructions labelled A–D, components of the arm are sequentially removed in order to reveal the components underneath. For colour examples of these reconstructions and additional detail concerning the morphology, see ref. 35.

structure. See Westbrook *et al.* (36) for examples of stereo pairs produced in this manner.

## 4.2 Fibre angle measurement

The measurement of fibre angles is often a necessity in the analysis of hydrostatic skeletons and is one of the reasons for obtaining serial sections. Measurements of fibre angle can be made easily on a microscope with a rotating stage equipped with a goniometer and ocular cross-hairs. One simply aligns the fibre of interest with one of the cross-hairs, notes the reading, and then rotates the stage until the cross-hair is aligned with a morphologically relevant line of reference; for instance, the longitudinal axis of a given structure. Often, the line of reference of the structure being studied is not easily defined, and it is more accurate to bisect the included angle between the right- and left-handed fibres. A variety of software packages for micro-computers with digitizing tablets are available that are also useful for making angular measurements. A tracing of the fibres can be made on a microscope equipped with a camera lucida and the tracing can then be digitized.

# 5. Functional analysis

## 5.1 Introduction

In addition to morphological analysis, research involving functional and ex-perimental analyses of hydrostatic skeletons is required. To date, relatively few experimental studies have been performed that serve to test the predic-tions of the biomechanical analyses of hydrostatic and muscular-hydrostatic skeletal support. In the following sections, some of the techniques that are useful in experimental analyses of these systems have been outlined with the hope that further work in this area will be encouraged.

## 5.2 Pressure measurement

The measurement of pressure is often critical to the understanding of hydrostatic skeletal support. A variety of pressure transducers are available commercially, many of which are remarkably small, sensitive and require only limited displacement of fluid. The transducers fall into several general categories:

- remote transducers that are connected via tubing (usually polyethylene) to a catheter implanted in the animal
- flush-mounted transducers that measure the pressure applied to a surface
- catheter tip transducers with the pressure-sensing diaphragm located at the tip of a catheter.

### 5.2.1 Remote transducers

The remote transducers are convenient to use because a variety of catheter bores and types can be used. Nevertheless, they are subject to movement artefacts due to vibrations of the tubing connecting the transducer to the catheter and to reduced amplitude and frequency response due to compliance of the tubing. For this reason, a remote transducer, and any other pressure transducer for that matter, should be not only calibrated, but its dynamic response should be determined. The dynamic response is most easily characterized using a 'pop test' in which one monitors the free resonant vibrations produced by a pressure transient applied at the end of the catheter. See Chapter 10 (Section 5.1) and Gabe (37) for details. An additional consideration in the use of remote transducers concerns potential blocking of the bore of the catheter with tissue debris. This is a particular problem in the study of muscular hydrostatic systems because fluid-filled cavities are not available for monitoring of pressure. A provision can be made for flushing of saline through the catheter. Alternatively a 'wick catheter' may be employed. A wick catheter is made by inserting a small piece of suture into the bore of the catheter (38). The suture prevents tissue from entering the bore of the catheter but allows pressure to be transmitted.

### 5.2.2 Flush-mounted transducers

Flush-mounted pressure transducers do not suffer from many of the movement artefact and frequency response problems of remote transducers because the sensing face of the transducer is exposed directly to the fluid under pressure. In addition, relatively little displacement of the sensing diaphragm occurs during measurement. Many of the flush-mounted transducers are miniaturized and could possibly be implanted in an animal.

### 5.2.3 Catheter tip transducers

The catheter tip pressure transducers also avoid some of the problems of remote transducers because the sensing face of the transducer is located on the tip of catheter (for example, Mikro-Tip Catheter Transducer, Millar Instruments, Houston, Texas, USA). They are convenient to use because a catheter placement needle can be used to insert the transducer and surgery is often not required. In addition, they are sealed and are thus convenient to use on experiments with marine organisms who in general seem to have remarkable aim when it comes to covering electronic devices with sea water!

## 5.3 Electromyography

Electromyography provides an important experimental approach to the testing of predictions concerning the pattern and timing of muscle activity. Electromyography using fine-wire bipolar electrodes has been used extensively in the study of vertebrate musculoskeletal systems. Chapter 8 in this volume by Gans provides details of electrode design, amplification, and recording techniques.

Electromyography has not been used extensively, however, in the analysis of the biomechanics of hydrostatic skeletal support. In part this reflects some unique difficulties associated with recording from the musculature of hydrostatic skeletons. For example, in a study of the fin musculature of the cuttlefish *Sepia officinalis* (22), the bipolar fine-wire electrodes used commonly for electromyographic recordings were tried initially but the thinness and flexibility of the fins made electrode placement difficult and unreliable. An alternative technique using pairs of individually implanted tungsten fine-wire electrodes was developed. This technique allows precise and controlled electrode placement in thin muscle layers, as in the cuttlefish fin. Since similar difficulties with bipolar fine-wire electrodes may be encountered in future work applying electromyography to the study of hydrostatic skeletons, the technique is described below.

### 5.3.1 Individually implanted tungsten electrodes for electromyography

One end of a piece of tungsten wire (0.25 mm diameter) is etched in a saturated solution of sodium nitrite using electrical current to form a sharp tip (*Figure 4*). It is convenient to control the current applied to the tungsten wire using a variable transformer and a carbon rod suspended in the sodium nitrate solution to complete the circuit. The most durable electrode tips are formed using high current initially and then reducing the current during the final stages of the etching. The electrodes are then dipped in orthophosphoric acid for cleaning and removal of deposits formed during the etching process. The tungsten wire is cut approximately 3–4 mm from the tip and a tight loop is formed. The stripped end of 0.075 mm Teflon-coated annealed stainless-steel wire is then tied to the loop and soldered. Immersion in orthophosphoric acid aids in soldering. Because of the difficulty of achieving a good solder joint between the stainless and tungsten wire, the joint is also painted with silver paint to ensure electrical continuity before being encapsulated in a

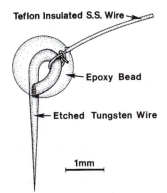

**Figure 4.** Diagram illustrating construction of one electrode of a pair of individually implanted tungsten electrodes for electromyography. Two such unipolar electrodes are implanted adjacently and are wired for bipolar recording. See text for details of fabrication.

small bead of epoxy resin. The electrode and attached wire is then dipped in insulator resin (3 coats) to ensure that both the stainless-steel wire at the solder joint and the tungsten electrode itself are insulated. The insulator resin is then removed from the tip of each electrode by scraping with a scalpel blade, forming a bared tip approximately 1 mm long. The wire leads of an electrode pair can be then glued together along their length and soldered to the appropriate connector.

The single electrodes of each bipolar electrode pair are then implanted individually. The bead of epoxy that encapsulates the connection between the stainless and tungsten wire provides a limit to the depth of implantation of the electrode. By varying the distance between the loop and the tip of the tungsten electrode, one can control precisely the depth of implantation of the electrode tip. The application of a thin layer of tissue adhesive (for example, Histoacryl, B. Braun Melsungen AG, Melsungen, W. Germany) over the electrodes, once inserted, helps to prevent accidental removal or changes in location of the electrode tips. It is also advisable to anchor the electrode leads to the animal so that any tension applied to the leads is borne by the anchor rather than by the electrodes themselves. A small plastic clamp capable of holding the leads without damaging the insulation is ideal.

Consult Chapter 8 for a discussion of considerations of preamplification, amplification, filtering, recording, digitization, and analysis of electromyographic signals.

## 5.4 Kinematics

Kinematic analysis is discussed in detail by Biewener and Full in Chapter 3 of this volume. The considerations of kinematic analysis of hydrostatic skeletons are similar to those discussed in Chapter 3. In many cases, kinematic data are collected on cine-film or video tape. Kinematic analysis using videotape or film is generally time-consuming and, as discussed by Biewener and Full, errors resulting from orientation and parallex complicate the analysis. Time can be saved with the use of a video motion analyser which provides a DC voltage that is proportional to the separation of two contrast boundaries on a horizontal line of a video image. See Chapters 10 and 11 and Gosline *et al.* (21) for a discussion of this approach. An alternative method for monitoring the movements of aquatic animals is described below that greatly reduces the time required for kinematic analysis and may also increase its accuracy and precision.

### 5.4.1 Electronic movement monitoring system for aquatic animals

The position of an animal or portion of an animal can be monitored by generating an electrical field in the tank and detecting the voltage measured by a sampling electrode at the location of interest on the animal. The circuit design for the system (*Figure 5*) was provided by Dr Douglas M. Neil,

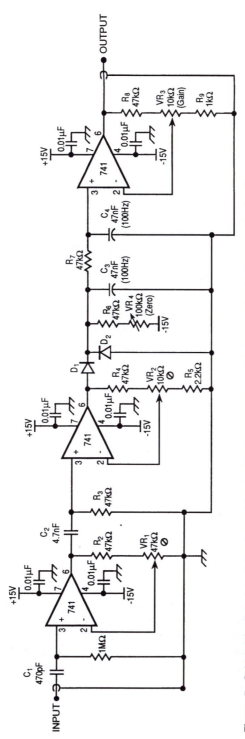

**Figure 5.** Schematic diagram of circuit used for monitoring movement in aquatic animals. See text for details.

229

Department of Zoology, Glasgow University, Scotland. The system is based on a circuit designed for the measurement of joint angles in arthropods (39) and was used to monitor the vertical position of the fin in a recent analysis of the muscular hydrostatic system of the fins of *Sepia officinalis* (22).

The movement monitor requires an experimental tank equipped with stainless steel plates on the bottom and at the surface of the water. The upper plate is connected to a function generator and a 40 kHz sine wave signal is supplied to it. The other plate is grounded. (For 30 cm × 25 cm plates separated by 12 cm in sea water, a 200 to 300 mV signal is required.) An electrode is made by simply removing 0.5–1.0 mm of insulation from 0.075 mm Teflon-coated annealed stainless-steel wire. The wire is then attached to the animal with tissue adhesive in such a way that the uninsulated end of the wire is located on the portion of the animal one wishes to monitor. The electrode is wired to the movement monitor circuit. The movement monitor rectifies, filters, and amplifies the input from the electrode and provides an output voltage that is proportional to the height of the electrode above the bottom of the tank. A number of electrodes and movement monitor circuits may be used in a single experimental tank and thus, the vertical position of several parts of the animal may be monitored simultaneously. The movement monitor signals can be recorded on separate channels simultaneous with other physiological recordings, providing direct correlation of movement with electromyograms, pressure recordings, etc.

# References

1. Chapman, G. (1958). *Biol. Rev.*, **33**, 338.
2. Alexander, R. McN. (1983). *Animal mechanics.*, Blackwell Scientific Publications, Oxford.
3. Chapman, G. (1975). *J. exp. Zool.*, **194**, 249.
4. Clark, R. B. (1967). *Dynamics of metazoan evolution.* Clarendon Press, Oxford.
5. Trueman, E. R. (1975). *The locomotion of soft-bodied animals.* Edward Arnold, London.
6. Wainwright, S. A., Biggs, W. D., Currey, J. D., and Gosline, J. M. (1976). *Mechanical design in organisms.* John Wiley, New York.
7. Kier, W. M. (1988). In *The mollusca* (ed. E. R. Trueman and M. R. Clarke), Vol. 11, pp. 211–52. Academic Press, New York.
8. Kier, W. M. and Smith, K. K. (1985). *Zool. J. Linn. Soc.*, **83**, 307.
9. Smith, K. K. and Kier, W. M. (1989). *Am. Scient.*, **77**, 28.
10. Kier, W. M. (1982). *J. Morphol.*, **172**, 179.
11. Kier, W. M. (1987). In *Nautilus: the biology and palaeobiology of a living fossil* (ed. W. B. Saunders and N. H. Landman), pp. 257–69. Plenum Press, New York.
12. Chapman, G. (1950). *J. exp. Biol.*, **27**, 29.
13. Clark, R. B. and Cowey, J. B. (1958). *J. exp. Biol.*, **35**, 731.
14. Harris, J. E. and Crofton, H. D. (1957). *J. exp. Biol.*, **34**, 116.
15. Seymour, M. K. (1983). *J. Zool.*, **199**, 287.

16. Woodley, J. D. (1967). *Symp. Zool. Soc. Lond.,* **20,** 75.
17. Woodley, J. D. (1980). In *Echinoderms: present and past* (ed. M. Jangoux), pp. 193–9. A. A. Balkema, Rotterdam.
18. Bone, Q., Pulsford, A., and Chubb, A. D. (1981). *J. Mar. Biol. Assoc. UK,* **61,** 327.
19. Gosline, J. M. and Shadwick, R. E. (1983). In *The mollusca* (ed. P. W. Hochachka), Vol. 1, pp. 371–98. Academic Press, New York.
20. Gosline, J. M. and Shadwick, R. E. (1983). *Can. J. Zool.,* **61,** 1421.
21. Gosline, J. M., Steeves, J. D., Harman, A. D., and Demont, M. E. (1983). *J. exp. Biol.,* **104,** 97.
22. Kier, W. M., Smith, K. K., and Miyan, J. A. (1989). *J. exp. Biol.,* **143,** 17.
23. Kier, W. M. (1989). *J. Zool. Lond.,* **217,** 23.
24. Glauert, A. M. (1975). *Practical methods in electron microscopy.* North-Holland, Amsterdam.
25. Bennett, H. S., Wyrick, A. D., Lee, S. W., and McNeil, J. H. (1976). *Stain Tech.,* **51,** 71.
26. Burns, W. A. and Bretschneider, A. (1981). *Thin is in: plastic embedding of tissue for light microscopy.* American Society of Clinical Pathologists, Chicago.
27. Steedman, H. F. (1957). *Nature,* **179,** 1345.
28. Pantin, C. F. A. (1946). *Notes on microscopical technique for zoologists.* Cambridge University Press, Cambridge.
29. Humason, G. L. (1979). *Animal tissue techniques* (4th edn). W. H. Freeman, San Francisco.
30. Cloney, R. A. and Florey, E. (1968). *Zeitschrift für Zellforschung,* **89,** 250.
31. Butler, J. K. (1984). *Stain Tech.* **59,** 315.
32. Latta, H. and Hartman, J. F. (1950). *Proc. Soc. exp. biol. Med.,* **74,** 436.
33. Aaron, B. B. and Gosline, J. M. (1980). *Nature,* **287,** 865.
34. Elias, H. (1971). *Science,* **174,** 993.
35. Kier, W. M. and Smith, A. M. (1990). *Biol. Bull.,* **178,** 126.
36. Westbrook, A. L., Haire, M. E., Kier, W. M., and Bollenbacher, W. E. (1991). *J. Morph.* (In press.)
37. Gabe, I. T. (1972). In *Cardiovascular fluid dynamics* (ed. D. H. Bergel), Vol. 1, pp. 11–50. Academic Press, New York.
38. Mubarek, S. J. (1976). *J. Bone Jt Surg. Am.,* **58,** 1016.
39. Marelli, J. D. and Hsiao, H. S. (1976). *Comp. Biochem. Physiol.,* **54A,** 121.

# 10

# Circulatory structure and mechanics

R. E. SHADWICK

## 1. Introduction: structural organization and fluid mechanics

This chapter describes techniques to investigate the structure and biomechanical characteristics of animal circulatory systems, with a focus on the viscoelastic properties of the arterial wall. These include quantification of connective tissue components, mechanical characterization of vascular tissue by *in vitro* and *in vivo* methods, calculation of vascular impedance and wave velocity from vessel elasticity, and assessing *in vivo* pressure–flow dynamics. For further details of haemodynamic principles and analysis the reader should consult the excellent monographs by McDonald (1), Milnor (2), and Fung (3).

## 1.1 Simple mechanical description of a circulation

In the simplest functional terms, animal circulatory systems are composed of a pumping organ and a set of conduit vessels carrying the blood or haemolymph. Circulatory pumps are powered by muscular contractions that generate oscillatory outflow. Ideally, however, flow to the exchange vessels (in closed systems) or sinuses (open systems) should be steady in order to reduce hydraulic power requirements and minimize the shear stresses on the vessel walls. The problem is overcome by the presence of elastic arterial vessels that are distended with blood when the heart contracts, and recoil when the heart is filling, thereby transforming the pulsatile output of the pump into a relatively steady flow at peripheral sites. Elastic compliance in the arterial system also reduces pulse pressure, hydraulic impedance, and pressure wave velocity which, in turn, influences other wave propagation characteristics. Thus, vascular elasticity is one of the most important haemodynamic determinants in a circulatory system. Non-linearity and viscoelasticity are also significant features of the walls of blood vessels.

### 1.1.1 Non-linear elasticity

Blood vessels typically have the property of elastic non-linearity. With increasing pressure a distensible vessel must become progressively stiffer (i.e. less compliant) in order to avoid elastic instability and rupture. This can be shown mathematically in several ways (4, 5, 6); the simplest being the expression of volume distensibility $D$ ($=dV/VdP$) which is inversely proportional to $((Eh/r) - P)$, where $E$ is the elastic modulus, $h$ is the vessel wall thickness, $r$ is the internal radius, and $P$ is the inflation pressure. As $P$ is increased, $r$ will increase and $h$ will decrease, assuming the length is unchanged (a reasonably good assumption for most cases). If $E$ remains constant, then $D$ becomes infinitely large as $P$ approaches $(Eh/r)$, and the vessel will rupture. This is avoided if $E$ is very large relative to $P$ (for example, a steel water pipe), but in that case the vessel would be too stiff to act as a pulse smoothing element. Alternatively, by having $E$ increase with distension, such that $(Eh/r)$ is always greater than $P$, the vessel can have large compliance at low pressures while being protected from unstable inflation at higher pressures. This situation has been universally observed in the arterial systems of the vertebrates and invertebrates studied so far (*Figure 1*). Typically, the range of normal physiological pressure in each animal corresponds with a range of rapidly increasing vessel wall stiffness. The non-linear behaviour of blood vessels reflects their

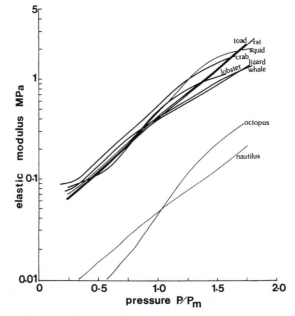

**Figure 1.** The elastic non-uniformity of blood vessels of various animals. Elastic modulus is plotted as a function of pressure ($P$) divided by the resting blood pressure ($P_m$) of each species.

composite structure that can be likened to a parallel arrangement of a rubbery elastic component with a stiff fibrous one (Section 2.1).

### 1.1.2 Viscoelasticity

Soft biological materials, like blood vessels, are not perfectly elastic because some degree of viscosity is always associated with their mechanical behaviour, i.e. they are viscoelastic. This means that the response to an external load is time-dependent and that some of the strain energy input is lost as heat by internal viscous processes (Section 3.4). In circulatory systems, pressure waves are transmitted as deformations of the vessel walls. Strain energy losses due to wall (and blood) viscosity provide a degree of amplitude attenuation that prevents reflected waves from resonating in the arterial system, and introduce frequency-dependent effects on wave velocity and hydraulic impedance [1].

## 1.2 Steady and pulsatile flow

### 1.2.1 The Poiseuille relation

An important but often misunderstood concept in cardiovascular fluid mechanics is that flow results not from a pressure, but from a pressure *gradient*. This principle is embodied in the simple Poiseuille description of steady flow in a rigid cylinder:

$$Q = \frac{\mathrm{d}P\pi r^4}{8\mu L} \tag{1}$$

where $Q$ is the volume flow rate, $\mu$ is the fluid viscosity, and $\mathrm{d}P$ is the pressure drop along a tube of length $L$ and radius $r$. This relation holds for laminar, non-turbulent steady flow in which the fluid viscosity gives rise to a parabolic velocity gradient across the vessel such that the velocity is zero at the luminal surface and maximal at mid-stream. Integration of the velocity across the luminal cross-section gives the mean volume flow rate $Q$. A non-parabolic velocity profile will occur in the entrance region of a tube, in turbulent flow, in fluids with viscosity anomalies, or in pulsatile flow.

Despite these limitations the Poiseuille relation is valuable because it provides a physical basis for the concept of vascular resistance, a measure of opposition to flow in a hydraulic system. By analogy to Ohm's law, resistance (**R**) is the ratio of mean pressure gradient to mean volume flow which, for the conditions of the Poiseuille relation, is:

$$\mathbf{R} = \frac{8\mu L}{\pi r^4} \tag{2}$$

### 1.2.2 Geometry of branching: Murray's law

Theoretical studies and experimental evidence from mammalian circulatory systems indicate that an optimum arrangement exists for the geometry of blood vessel branching in a flow network to minimize the energetic cost of conducting the fluid. A cost function that includes the power required to move the blood and the metabolic cost of maintaining the vascular tissue was formulated by Murray (7). By using Poiseuille's relation, he predicted that the optimal radius for any vessel (and thus for the collection of vessels in a system) is proportional to $Q^{1/3}$. If flow is conserved, this means that $r^3$ for the parent vessel equals the sum of $r^3$ for its daughter vessels. In other words, at each level of branching in the optimum vascular system the total flow will be carried by a set of vessels, the sum of whose radii cubed is a constant. Consequential predictions of Murray's law are that the total cross-sectional area and resistance of each level of branching increase as $r^{-1}$, the flow velocity increases as $r^1$, and the shear stress on the vessel wall is independent of $r$. Data from anatomical studies of mammals tend to support this prediction. No similar data are yet available for the circulatory systems of other vertebrates or invertebrates.

### 1.2.3 Pulsatile flow

Of the models developed to describe dynamic $P$–$Q$ relations, the classic two-element Windkessel is the simplest. This consists of an elastic chamber, or Windkessel, representing the elastic arteries, in series with a resistor representing the peripheral vascular beds (*Figure 2*). An alternative representation is by an electrical analogue circuit consisting of a capacitor in parallel with a resistor. The Windkessel chamber is distended with blood during systolic ejection and discharges blood through the resistance by elastic recoil during diastole. The intermittent flow from the heart is thereby transformed to a

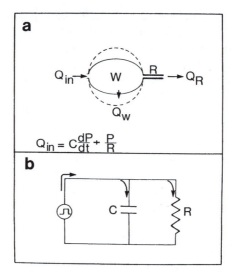

**Figure 2.** (a) A representation of the Windkessel hydraulic model of a circulation. (b) An electrical analogue in which the elastic chamber is represented by a capacitor (C), charging and discharging through a peripheral resistor (R).

quasi-steady flow in the periphery. The total flow into the system ($Q_{in}$) is the sum of the rate of volume change in the Windkessel ($Q_W$) and the simultaneous flow through the resistance ($Q_R$). The pulsatile component $Q = (dP/dT)(dV/dP)$, where $dV/dP = C$, the compliance of the chamber. The steady component $Q_R = P/R$. During diastolic discharge the pressure will decline exponentially from the end-systolic value $P_s$ as:

$$P(t) = P_s\, e^{(-t/RC)} \tag{3}$$

The time-constant $RC$ characterizes the combined effect of the elasticity and resistance of the system. $RC$ can be estimated from the diastolic part of a pressure waveform recorded *in vivo* as:

$$RC = T_d/\ln(P_s/P_d) \tag{4}$$

where $T_d$ is the duration of diastole and $P_d$ is the final diastolic pressure. An alternative calculation of $RC$, based on integrating the pressure curve is given by Yin *et al.* (8). With the time-constant and an estimate of $R$ from mean blood pressure and cardiac output data, the arterial compliance $C$ may be calculated. On the other hand if $C$ and $P$ are known, then $R$ and cardiac stroke volume can be calculated.

The Windkessel model is simplistic in that it assumes that the pressure pulse is transmitted instantaneously throughout the system and that no interactions occur between successive pulses, via reflections. While these assumptions are nearly met in the circulatory systems of poikilotherms with low heart rates (9, 10) more complex models are required to account for the transmission properties of mammalian systems (see refs. 1, 2, and 11).

A parameter of importance in formulating pulsatile models is the non-dimensional term referred to as Womersley's number $\alpha$:

$$\alpha = r(\omega\rho/\mu)^{1/2} \tag{9}$$

where $\rho$ is the fluid density and $\omega$ is the angular frequency ($\omega = 2\pi f$). This expresses the relative importance of inertial to viscous components in pulsatile flow, as does Reynolds number for steady flow (see Chapter 11). For $\alpha \ll 1$, pulsatile flow is dominated by viscous forces and the flow will be in phase with the oscillatory pressure gradient. With increasing values of $\alpha$, transient inertial forces become dominant. The velocity profile becomes blunted and the flow pulse decreases in amplitude and lags the pressure gradient by an increasing phase angle, up to a limiting value of 90 degrees for $\alpha \gg 10$. Formulae for calculating pulsatile flow from pressure gradients can be found in refs. 1 and 2.

# 2. Morphological and biochemical techniques

## 2.1 The artery wall composite structure

Arteries are typically composite structures that consist of cellular and connective tissue components, primarily smooth muscle, elastic tissue, and collagen.

Collagen is a relatively stiff and inextensible fibrous protein that provides a reinforcing effect to most soft biological tissues. Collagen fibres typically have a tensile strength of greater than 100 MPa, and an elastic modulus (stiffness) of 1–2 GPa (12). On the other hand, elastic tissues (elastin in the vertebrates and analogous proteins in invertebrates) are highly extensible rubber-like proteins whose elastic modulus is only 1/1000 that of collagen (13). Clearly, the contribution of each of these constituents to the mechanical properties of the composite will depend largely on their relative abundance and orientation, and on the level of inflation pressure. Thus, the composition and the fibrous architecture should be studied in parallel with vascular wall mechanics.

In a simplistic approach, the elasticity of the vessel wall can be calculated by apportioning the properties of the two main elements based on their volume fractions, and assuming that these elements act in parallel and the vessel wall is a linear material. Thus:

$$E_v = E_e v_e + E_c v_c \qquad (6)$$

where $E$ denotes an elastic modulus, $v$ is a % dry weight, and the subscripts v, e, and c denote the vessel wall, the elastic component, and the collagen, respectively. The weakness of this model is that it does not account for the orientation and degree of loading of the elements. Cox (14) proposed a better formula that allows the fraction of collagen fibres being loaded ($f_c$) to vary as a function of pressure or distension, and thereby account for the non-linear wall properties:

$$E_v = E_e v_e + f_c E_c v_c. \qquad (7)$$

If the vessel wall properties and composition are known, equation (7) can then be used to solve for $f_c$ as a function of pressure.

## 2.2 Structural studies

Histological or electron microscopical examination of transverse and longitudinal sections of the artery wall are useful in determining the orientation of fibrous elements including smooth muscle, elastic fibres, and collagen, as well as providing a qualitative assessment of their relative abundance.

### 2.2.1 Recommended staining procedures for connective tissues of artery wall

Conventional histo-staining techniques that identify elastic tissue in vertebrate tissues are not all suited for staining analogous structures in invertebrates. This has led to some discrepancy in identifying presumptive elastic tissues in some invertebrates. We find that the aldehyde–fuchsin method, described by Humason (15), works well for both groups of animals (*Figure 3*). The procedure for this stain requires the tissue to be pre-oxidized with acidified potassium permanganate, and it is currently believed that the N-ethylated derivatives of pararosaniline in the staining mixture react with

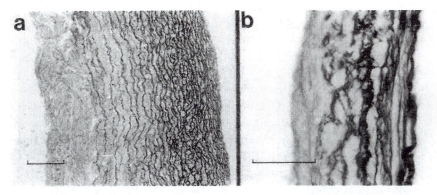

**Figure 3.** Staining of elastic fibres by aldehyde–fuchsin in (**a**) *Nautilus* aorta, and (**b**) toad aorta. The luminal surface is to the right in both. Scale bars are 100 μm in **a** and 50 μm in **b**.

aldehyde groups on the protein, presumably generated by the oxidation (16). Note, however, that:

(a) Aldehyde–fuchsin must be made fresh from a reaction involving paraldehyde and basic fuchsin (pararosaniline, C.I. 42500). Some stains sold as basic fuchsin are not pure pararosanaline and these should be avoided.

(b) The major disadvantage of this technique is that paraldehyde is now a controlled substance, due to its toxicity, making it somewhat expensive and inconvenient to use.

(c) An alternative staining reaction, called spirit-blue (17), may be preferred. This procedure also requires permanganate oxidation of the tissue, followed by staining with alcohol soluble aniline blue. The results appear to be similar to those using aldehyde–fuchsin (17).

(d) In both cases counterstaining with van Gieson's picro-fuchsin allows the collagen and muscle components to be distinguished.

## 2.3 Biochemical analyses

### 2.3.1 Vertebrates

Typically, elastin and collagen in vertebrate tissues are quantified by methods that take advantage of their unique biochemical properties. Mature elastin is a highly stable, covalently crosslinked protein that is also very hydrophobic. It is isolated from arterial tissue as the insoluble residue after extraction procedures that remove all other components. Collagen is a relatively insoluble, crosslinked protein that is quantified by determining the tissue content of hydroxyproline (HYP), an amino acid not normally found in other proteins (it is, however, found as a minor component of elastin in some species). The following protocol represents the simplest approach to elastin and collagen determination. Greater detail and alternative methods are found in refs. 18–20.

**Protocol 1.** Quantifying the collagen and elastin of the vertebrate
artery wall

*Elastin*

1. For each site of interest, excise the vascular tissue, rinse in cold, buffered saline, and chop finely with a razor blade.

2. Blot well and weigh a small portion of the sample. Freeze-dry and weigh again to determine tissue water content. Save this sample for collagen analysis (see below).

3. Blot well and weigh the remaining tissue. Calculate the dry weight from step 2. Homogenize or mince in a small volume of 0.1 N NaOH.

4. Add nine volumes of 0.1 N NaOH and immerse sample container in a boiling water bath for 45 min. Suitable containers would be Corning glass or polypropylene culture tubes with screw caps.

5. When cool, centrifuge the sample. Wash the pellet twice in cold distilled water (DW), freeze-dry and weigh the residue. This will be the elastin fraction. Hydrolyse this residue (see below) and subject it to amino acid analysis or, at least, an HYP assay.[a]

6. Calculate the elastin content as a percentage of the total tissue dry weight.

*Collagen*

1. Take 1 to 5 mg of the freeze-dried artery wall from step 2 above, and add 0.5 to 1 ml of 6 N HCl in a glass tube that can be sealed under vacuum. Hydrolyse the sample by heating to 110°C for 24 h.[b]

2. Dry the hydrolysate by flash evaporation or dilute and freeze-dry using a cold trap.

3. Dissolve in the appropriate buffer and load a measured aliquot on an amino acid analyser to determine the content of HYP. Calculate the collagen content of the tissue dry weight, based on the assumption that vertebrate fibrous collagen has 110 HYP residues per 1000 amino acids (representing 14% of the collagen dry weight).

4. Correct to account for the HYP contributed by the elastin in the tissue. To do this the dry weight percentage of elastin and its HYP content must be known. This correction may be significant if the tissue contains a large proportion of elastin, particularly in lower vertebrates where the HYP content of elastin may approach 3% (21).

[a] If an amino acid analyser is unavailable, HYP can be quantified by a manual colorimetric assay of hydrolysates following the procedure described by Berg (22). This technique is useful for HYP concentrations down to 0.5 μg/ml.

[b] Alternatively, hydrolyse in a mixture of 7 N HCl, 10% trifluoroacetic acid and 0.1% phenol, and heat for 45 min at 155°C (23).

## 2.3.2 Invertebrates

The arteries of invertebrates are typically composite materials that have close structural and functional analogy to those of vertebrates, i.e. an extensible elastic tissue is reinforced by stiff collagen fibres (24–26). Quantification of collagen in invertebrate vascular tissues can be done according to *Protocol 1*. Elastic proteins are expected to be stably crosslinked, but not necessarily as hydrophobic and resistant to NaOH as elastin. Very little is known about the chemistry of vascular elastomers from invertebrates. *Protocol 2* is offered as a procedure for their preparation, based on experience in my laboratory.

---

**Protocol 2.** Isolation of presumptive elastic tissue from invertebrate blood vessels

1. Follow steps 1–3 from *Protocol 1*, but substitute DW for the 0.1 N NaOH.

2. Transfer sample to a sealable glass container, add 20 volumes of DW and autoclave at 1 atm for 45 min. Centrifuge and repeat autoclaving with fresh DW until no protein is detected in the supernatant (at least 3 times).[a]

3. Freeze-dry and weigh the residue. This can be taken tentatively to be the insoluble elastic component.[b] Hydrolyse this residue and subject it to amino acid analysis or, at least, a HYP assay.

[a] Mix equal volumes of the supernatant and 10% trichloroacetic acid. If protein is present it will form a cloudy precipitate.

[b] This material should be subjected to further biochemical study to assess its purity and structural/immuno-histochemical study to establish its location in the connective tissue of the native vessel.

---

# 3. Mechanical testing of vascular tissue

This may be carried out in several ways, the most common being uniaxial tests on rings and strips, biaxial inflation tests of vessel segments, and *in vivo* determinations using the natural pressure pulses generated by the heart. Examples of each are given below. The main objective is to measure the mechanical response of the tissue to some systematic deformation, i.e. to determine the stress–strain relationship under conditions that have physiological meaning. Detailed descriptions of various testing equipment and procedures are available (25–34). Generally, if the passive elastic properties are to be investigated, the vessel can be tested immediately or after storage for several hours in suitable saline at 4°C. If the active properties of the muscle component are also of interest keep the vessel in a buffered saline containing glucose and aerated with a mixture of 95% $O_2$ and 5% $CO_2$, and test immediately. For details of investigations of the contribution of muscle to the mechanical properties of arteries see (34, 35).

A common phenomenon observed during *in vitro* tests of vascular tissue, as well as other soft biological tissues and man-made polymers, is the 'conditioning' effect of repetitive cycles of deformation (36, 37). Initially, successive pressure–volume or force–extension curves shift progressively to the right (that is, increased volume or extension), but the difference between each cycle diminishes until stable loading and unloading curves are obtained. It appears that a period of adjustment of the internal structure to renewed stress is necessary following excision from the animal (36). Thus, no conditioning period is required when mechanical properties are measured *in vivo*.

## 3.1 Inflation tests *in vitro*

This is the method of choice for isolated vessels because it is biaxial and comes closest to mimicking the *in vivo* situation. Carefully excise the blood vessel post-mortem, after ligating the branch vessels and noting the *in vivo* length of the segment to be used. Excision is usually accompanied (at least in the vertebrates) with a significant elastic retraction resulting in a resting length that is much shorter than *in vivo*. It will probably be necessary to re-establish the *in vivo* length for testing, or at least to know how much shortening has occurred. The purpose of this test is to measure the deformation of the artery wall at different pressures, and subject these data to an engineering type of analysis of material properties. Tests to failure are rarely done on arteries, as this is not usually of physiological interest.

The assumptions underlying this approach are that the blood vessel is cylindrical, the wall is incompressible and homogeneous, and the stresses occur in orthogonal directions (34).

---

**Protocol 3.** Inflation of blood vessel segments at constant length
(*Figure 4*)

1. Cannulate the vessel segment by tying both ends over tubular connectors, made from metal cylinders, syringe needles, or PE-tubing, depending on the vessel diameter.

2. Mount the vessel segment horizontally in a test chamber with the appropriate saline by coupling the end connectors to an arrangement of three-way stopcock valves held in position by clamps.[a]

3. Connect a linear perfusion pump and a pressure transducer to the stopcock array to provide continuous inflation of the vessel and simultaneous measurement of pressure.[b] Perfuse the artery to clear air bubbles and check for leaks. Close the outflow valve.

4. Measure the diameter at zero pressure. Stretch the vessel to the predetermined *in vivo* length. The axial force required to hold the artery at this length can be measured if the vessel is tethered via a force transducer. Measure the new external dimensions.

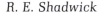

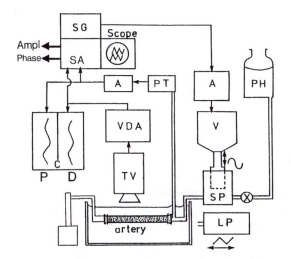

**Figure 4.** A schematic of an inflation apparatus. The artery segment is cannulated and held in a saline bath, and inflated either by a linear pump (LP), or a sinusoidal pump (SP) driven by a vibrator (V). The pressure inside the vessel is measured by a pressure transducer (PT) and the external diameter by a video system consisting of a TV camera and a video dimension analyser (VDA). Quasi-static pressure (P) and diameter (D) signals are digitized and recorded on a micro-computer (C). Signals from dynamic tests are played into a digital spectrum analyser (SA) which contains a signal generator (SG) to power the vibrator. A = amplifier; PH = pressure head.

5. Inflate the vessel, recording pressure and diameter (and axial force if required) simultaneously.[c] Repeat until stable inflation–deflation cycles are obtained. This may be repeated at different inflation rates, depending on the capability of the pump and measuring devices.

6. Following the tests, section the vessel transversely (use a freezing microtome if possible) and make measurements of the resting wall thickness.

7. Alternatively, calculate the unstressed wall thickness from the wall volume of a known length of the vessel. This can be obtained by Archimedes' principle; place a container of water on a weighing balance and suspend the vessel segment into it. The increase in weight represents the volume of water displaced by the specimen.

[a] We use magnetic base clamps and set the test apparatus on a large steel plate.
[b] If a roller-type pump is used, connect an air chamber in series to smooth the flow.
[c] See Section 5 for details of pressure and dimension measurement.

### 3.1.1 Inflation of an untethered vessel

This approach is appropriate for vascular tissue from soft-bodied animals in which the blood vessels are not tethered to a fixed length *in vivo*. The procedure is the same as above except that the vessel is not stretched longitudinally, nor restricted from lengthening when inflated. It is then necessary to measure changes in length as well as diameter and inflation pressure.

## 3.2 Uniaxial tests *in vitro*

This method is less appropriate than biaxial tests but is useful because it is relatively simple to conduct. A good description of uniaxial testing of dog arteries is given by Attinger (38).

To test the longitudinal properties, cannulate a vessel segment as above and then stretch it by gripping the tubular connectors while holding the specimen in a saline bath. This will allow fluid to move out of the lumen as the diameter decreases with longitudinal extension, keeping the internal pressure at zero (*Figure 5a*). If the ligatures are tight, slippage should not be a problem. For circumferential tests, cut a ring from the vessel and mount it over U- or L-shaped pins that can be gripped in a test apparatus (*Figure 5b*).

In either case, mount the specimen in a tensile test machine, such as an Instron, and measure the force and extension continuously as the specimen is cyclically stretched at relatively slow rates. Alternatively, extend the specimen stepwise by pulling one end with a micrometer device and measuring force with an in-series transducer, or load it with a series of increasing dead weights and measure extension with a displacement transducer or rule. If possible, restrict extension measurements to the central region of the sample, because local stress concentrations in the tissue at the clamps may cause strain artefacts. Determine the cross-sectional area by measuring wall thickness or volume, as above.

## 3.3 Data analysis

### 3.3.1 Uniaxial tests

Force and deformation data must be normalized by the sample dimensions to values of stress ($\sigma$, in Pa) and strain ($\varepsilon$) in order to assess the material properties of the tissue. A plot of $\sigma$ vs. $\varepsilon$ characterizes the passive mechanical properties, and the slope of this curve is a measure of the stiffness or elastic modulus ($E$) of the tissue. Biological materials like blood vessels typically have non-linear stress–strain curves, in which the elastic modulus varies with the level of strain. It may be defined, for each strain, as the tangent of the stress–strain curve at that strain. Thus:

$$E = \Delta\sigma/\Delta\varepsilon. \tag{8}$$

This simple relation applies only to uniaxial deformation. Thus, force and extension data from longitudinal or circumferential tensile tests, normalized to stress and strain, can be used to calculate the elastic modulus in those directions. For longitudinal tests stress and strain are calculated as:

$$\sigma_L = F/A = F/[\pi(R^2 - r^2)] \tag{9}$$

$$\varepsilon_L = \Delta L/L_0 \tag{10}$$

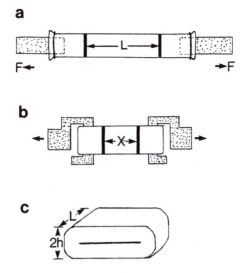

**Figure 5.** Uniaxial testing schematic. In (**a**) a vessel segment is stretched longitudinally and the extension (L) is measured between two surface markers. In (**b**) a ring is stretched circumferentially and extension (X) is measured between two surface markers. (**c**) shows the terms used to describe the dimensions of the ring specimen.

where $R$ is the external radius, $r$ is the internal radius, $L$ is the stretched length and $L_0$ is the initial length. Assuming that the artery wall is incompressible, $A$ can be calculated at each length from the initial dimensions [that is, $(R^2 - r^2)$ decreases inversely with increasing $\varepsilon_L$]. For circumferential tests of rings, $\sigma_C$ is calculated as $F/2hL$, and $\varepsilon_C$ is determined from uniaxial extension as $\Delta X/X_0$ (*Figure 5*).

### 3.3.2 Biaxial inflation data

*i. Stress–strain relations*

Inflation of a blood vessel *in vitro*, as *in vivo*, results in simultaneous strains in at least three orthogonal directions, and the stress analysis becomes more complex than for uniaxial extensions. For simplicity we consider the circumferential and longitudinal tensile forces, and neglect the relatively small radial compressive forces. For example, a thick-walled elastic tube distended by pressure $P$ (*Figure 6*) will experience a circumferential stress given by:

$$\sigma_C = Pr/h. \tag{11}$$

The circumferential strain can be calculated from changes in the external, internal, or preferably, the mid-wall radius:

$$\varepsilon_C = \Delta \hat{R}/\hat{R}_0 \tag{12}$$

where $\hat{R} = (R + r)/2$. The longitudinal stress imposed by the pressure is:

$$\sigma_L = Pr^2/2\hat{R}h. \tag{13}$$

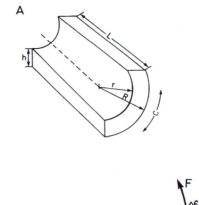

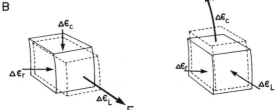

**Figure 6. A**: Diagram to show the terms used for the dimensions of a vessel segment. **B**: Illustrations of deformations involved in Poisson's ratios. A uniaxial force (*F*) in the longitudinal direction causes narrowing in the circumferential and radial directions. A uniaxial force in the circumferential direction causes a reduction in the length and a narrowing in the radial direction.

As $R/h$ increases, $r$ approaches $R$ and $\sigma_L$ approaches $\sigma_C/2$ (the case for a thin wall tube). Longitudinal strain is given by equation (10). If the artery is inflated with no longitudinal tethering then $\hat{R}_0$ and $L_0$ are the unstressed vessel dimensions.

If the artery is stretched and held at a constant length for testing, then $\Delta\varepsilon_L$ remains zero and it is appropriate to refer to the stretched, unpressurized radius in equation (12) to determine $\varepsilon_C$. In this case an additional component of stress results from the tethering force [see equation (9)], so that the total longitudinal stress will be:

$$\sigma_L = Pr^2/2\hat{R}h + F_L/[(\pi(R^2 - r^2)]. \tag{14}$$

Values of $r$ and $h$ can be calculated at each pressure by knowing $L$ and the wall volume $[V_w = \pi(R^2 - r^2)L]$, which is assumed to be constant. Each pressure increment produces a circumferential strain that depends on the stress and elastic modulus in that direction, but which is also influenced by the strain increment that occurs simultaneously in the longitudinal direction. The interaction between orthogonal strains in an elastic material is given by Poisson's ratio ($v$). At least two Poisson's ratios are necessary to describe the relationship between $\varepsilon_C$ and $\varepsilon_L$ in an artery:

$$v_{CL} = -\Delta\varepsilon_C/\Delta\varepsilon_L \tag{15}$$

$$v_{LC} = -\Delta\varepsilon_L/\Delta\varepsilon_C \tag{16}$$

where $v_{CL}$ is the ratio of the decrease in circumference that would accompany a uniaxial longitudinal strain and $v_{LC}$ is the ratio of the decrease in length that would accompany a uniaxial increase in circumference (*Figure 6*). Values of $v$ normally vary between 0 and 1.0. If the artery wall is anisotropic (i.e. the longitudinal and circumferential elastic moduli are not equal) the Poisson's ratios are symmetrically related as:

$$v_{LC}/E_C = v_{CL}/E_L. \tag{17}$$

When an artery is inflated the actual changes in circumferential and longitudinal dimensions depend on the Poisson's ratios and elastic moduli. If the radial stresses are ignored, the biaxial stress–strain data for each pressure increment can be analysed according to the following relation (34):

$$\Delta\varepsilon_C = (\Delta\sigma_C/E_C) - v_{CL}(\Delta\sigma_L/E_L) \tag{18}$$

$$\Delta\varepsilon_L = (\Delta\sigma_L/E_L) - v_{LC}(\Delta\sigma_C/E_C). \tag{19}$$

Comparison with equation (8) shows that the biaxial strains will be less than uniaxial ones by an amount that depends on the properties in the other direction [that is, the second term on the right side of equations (18) and (19)]. Substituting from equation (17) into equation (19) yields:

$$\Delta\varepsilon_L = (\Delta\sigma_L/E_L) - v_{CL}(\Delta\sigma_C/E_L) \tag{20}$$

which allows these relations [equations (18) and (20)] to be given in terms of only one Poisson's ratio and the elastic moduli.

## ii. Untethered, anisotropic vessel

If $v_{CL}$ can be measured directly in uniaxial tests, then $E_L$ and $E_C$ are solved from equations (20) and (18). Alternatively, if $\Delta\varepsilon_L$ is small compared to $\Delta\varepsilon_C$, $E_L$ may be determined from uniaxial extensions, and substituted into equation (20) together with the biaxial data to obtain $v$, and then $E_C$ from equation (18). This method was used in studies of the aorta of various cephalopods (25, 39, 40), where $v$ was found to range from about 0.3 to 0.5.

## iii. Tethered, anisotropic vessel

This is simplified by the fact that $\Delta\varepsilon_L = 0$, so equations (20) and (18) yield, respectively:

$$v_{CL} = \Delta\sigma_L/\Delta\sigma_C \tag{21}$$

$$\Delta\varepsilon_C = (\Delta\sigma_C/E_C) - (\Delta\sigma_L/\Delta\sigma_C)(\Delta\sigma_L/E_L). \tag{22}$$

To use equation (22), the biaxial data must include measurements of the tethering force at each pressure [i.e. equation (14) must be used for $\sigma_L$].

Then, substituting a value for $E_L$ at the tethered length (determined uniaxially) into equation (22) allows the calculation of $E_C$ for each pressure increment.

### iv. Tethered, isotropic vessel

A further simplification may be introduced by assuming that the vessel wall is isotropic; that is, $E_C = E_L$ and $v_{CL} = 0.5$. This makes the measurement of tethering forces and $E_L$ unnecessary. Equations (18) and (20) reduce to give:

$$E_C = (\Delta\sigma_C/\Delta\varepsilon_C)\,(1 - v_{CL}^2), \tag{23}$$

or, $E_C$ is 0.75 of the slope of the circumferential stress–strain curve. Generally the walls of arteries are not isotropic, but equation (23) is often used because of simplicity, particularly with *in vivo* pressure–diameter data (see Section 3.5).

### v. Incremental formulae

Bergel (27) defined an incremental elastic modulus ($E_{inc}$) that is often used in studies of arterial elasticity. This is based on pressure and diameter changes of tethered vessels, assumed to be isotropic, and an incremental strain $\varepsilon_{inc}$ that is defined as the change in radius divided by the average radius during the change, $R'$ (i.e. $\varepsilon_{inc} = \Delta R/R'$). Bergel's formula was expressed directly in terms of pressure and radius changes, rather than stress and strain:

$$E_{inc} = \Delta P/\Delta R\,(r^2 R/R^2 - r^2)\,2\,(1 - v_{CL}^2). \tag{24}$$

If $R/h$ is small, then equation (24) can be simplified to:

$$E_{inc} = (\Delta\sigma_C/\varepsilon_{inc})\,(1 - v_{CL}^2). \tag{25}$$

Note that equations (23) and (25) differ only by the factor $(1 + \varepsilon)$; that is, $E_{inc} = (1 + \varepsilon)E_C$, since $R' = R_0\,(1 + \varepsilon)$.

## 3.4 Viscoelasticity: measuring time-dependent properties

### 3.4.1 *In vitro* methods

#### i. Hysteresis

The simplest demonstration of viscoelasticity in vascular tissue comes from cyclic pressure–volume or force–extension curves. After conditioning cycles are complete, the unloading curve will remain below the loading curve (*Figure 7*). The strain energy required to deform the vessel is represented by the area under the loading curve, while the energy released by elastic recoil is proportional to the area under the unloading curve. The area between these curves represents the energy lost by viscous damping. The relative importance of the viscous component can be expressed by the mechanical hysteresis, defined as the ratio of the energy lost to the total energy absorbed during one cycle. This can be measured simply by cutting out and weighing the areas from recorded traces, or calculated by integration of digitally recorded data. Typical values of hysteresis for arteries from vertebrates and invertebrates range from 10 to 20%.

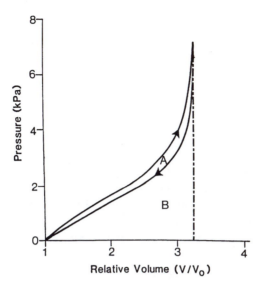

**Figure 7.** A typical inflation curve for an artery (dorsal aorta of a toad), showing the pressure–volume curves for loading and unloading (arrows). The hysteresis is calculated as the area A (= energy lost) divided by the area A + B (= total energy input.) (From C. A. Gibbons and R. E. Shadwick, unpublished.)

## ii. Dynamic testing

To measure viscoelastic properties at deformation rates that match or exceed those normally imposed by the heart, the method of forced vibrations can be used. This technique requires:

(a) A pump that can load the specimen sinusoidally.

(b) Pressure and diameter transducers that can respond adequately to the frequencies used (Section 5).

(c) An instrument that can detect the amplitude and phase of the transduced signals (for example, spectrum analyser or a micro-computer with wave-form analysis software. A good selection of these items is available from Perx Engineering Inc., Corte Madera, California, USA).

---

**Protocol 4.** Dynamic test *in vitro*

1. Cannulate an artery segment as in *Protocol 3*.

2. Inflate by a static pressure while imposing small volume oscillations at different frequencies with a pump, usually driven by an electromagnetic vibrator (such as those made by Ling Dynamic Systems, UK) or eccentric cam motor. This simulates the *in vivo* loading of the artery.

**Protocol 4.** *Continued*

3. Measure the amplitudes and phase difference of the pressure and diameter responses at each frequency. Examples of this type of experiment are described in (25, 28). The viscoelasticity will cause the pressure (and stress) oscillation to be phase-shifted ahead of the diameter (and strain) oscillation by an angle δ. The magnitude of δ indicates the importance of viscous damping at each frequency (see refs. 37 and 41 for theoretical details, and refs. 2 and 42 for discussion of the application to vascular mechanics).

4. Calculate the dynamic tangent elastic modulus, $E^*$, from the pressure and diameter oscillations as described in Section 3.3. If the vessel is tethered, or if the diameter oscillations are sufficiently small that $\Delta\varepsilon_L$ is negligible, the simpler relations in equations (21–23) may be used.

---

$E^*$ can be resolved into the storage modulus ($E'$) and the loss modulus ($E''$) for each frequency and mean pressure in the following way:

$$E' = E^* \cos \delta \quad \text{and} \quad E'' = E^* \sin \delta. \tag{26, 27}$$

$E'$ is derived from the component of stress that is in phase with the strain, and is proportional to the energy stored elastically in each cycle. $E''$ is derived from the out-of-phase component of stress and is proportional to the viscous energy lost per cycle. The ratio $E''/E' = \tan \delta$ and is called the viscous damping or phase loss. This can be related to hysteresis as ($2\pi \tan \delta$) when δ is small. Dynamic data allow one to assess how the stiffness and elastic storage properties vary at different mean pressures and pulsatile frequencies. $E'$ can also be used to calculate haemodynamic parameters, such as pulse wave velocity and characteristic impedance (Section 4).

*iii. Creep and stress relaxation*

Creep (the increase in strain under constant stress) and stress relaxation (the decay in stress at constant strain) have also been used to characterize the time-dependent properties of vascular tissue (38, 43–45). These are transient methods, generally suited to the study of viscoelasticity on the order of 1 sec or longer, and can be interrelated with forced vibration tests qualitatively or mathematically (37, 41). The results are expressed as creep rate, creep compliance, or relaxation modulus, and can be used to evaluate viscoelastic models of the tissue.

There are a number of technical difficulties that make these methods less appealing than using forced vibrations. For intact vessel segments, stress relaxation is measured from the decay in pressure following a rapid volume increase. Fluid leakage as well as viscoelasticity will cause a gradual decrease in pressure, and it is difficult to separate the two effects. Creep is measured as the increase in diameter while the vessel is held at a constant inflation pressure, a seemingly simple experiment. However, as the diameter increases, the wall stress will also increase [equation (11)] rather than remaining constant. This will result in creep data that can not be analysed in terms of simple viscoelastic models.

### 3.4.2 *In vivo* methods

The most appropriate measurements of the elastic properties of a blood vessel are made by *in vivo* techniques. Pressure and diameter pulsations are recorded simultaneously in a living animal, and the dynamic elastic modulus is then calculated for the range of pressure and frequency normally generated by the heart. This range may be extended somewhat by altering the blood pressure or heart rate experimentally. While *in vivo* measurements are desirable, they are not essential in order to relate the mechanical properties of the vessel to its haemodynamic function. Several previous studies have confirmed that the elastic properties of an artery measured *in vitro* are essentially the same as those *in vivo*.

*i. Requirements for in vivo measurements*

(a) The animal must be anaesthetized and the blood vessel of interest exposed. A cannula from a pressure transducer is introduced, preferably via a side branch to record pulsatile blood pressure.

(b) A diameter-measuring device must be applied to record radial pulsations as close as possible to the position of pressure measurement. If this is a mechanical, optical, or video instrument then recordings will be possible only while the artery is exposed. If an ultrasound technique is available, this may allow pressure and diameter pulses to be measured chronically after the animal has recovered. Ultimately the wall thickness must be determined for the range of pressure studied.

(c) Data should be recorded digitally if possible, or at least in a form that can be digitized later. Sampling rates should be at least 10 times greater than the highest frequency of importance in the signal.

(d) The pressure and diameter transducers must be calibrated dynamically and phase and amplitude corrections applied to the data if necessary (see Section 5).

(e) The calculation of circumferential stress, strain, and dynamic elastic modulus will be similar to the analysis of *in vitro* dynamic tests, once the complex $P$ and $D$ waveforms have been resolved into their sinusoidal components by the use of Fourier analysis. This provides the important quantities of amplitude and phase for each component (see *Figure 8*).

*ii. The use of Fourier analysis*

It is convenient to deal with periodic events in terms of sinusoidal waves each defined by a frequency, amplitude, and phase shift. For example, $M \cos(\omega t - \phi)$

251

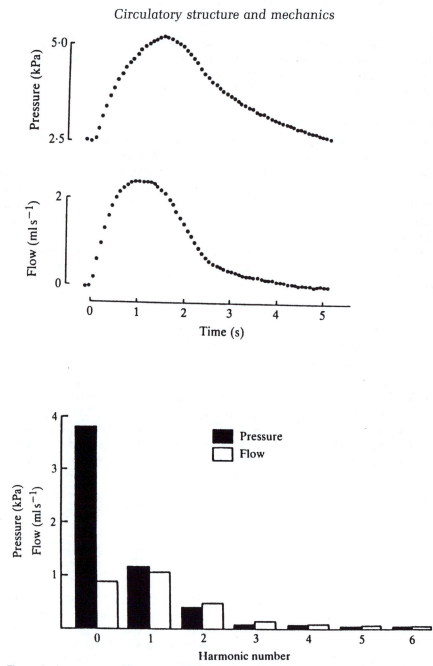

**Figure 8.** An example of Fourier analysis of digitized pressure and flow waves recorded in the aorta of an octopus. The amplitude of each pressure and flow harmonic is shown in the histogram. The zero harmonic represents the mean value. The first harmonic occurs at the fundamental frequency while the higher harmonics are at multiples of this frequency. (From ref. 10.)

is a function that has amplitude $M$ (also called the modulus), circular frequency $\omega$, in radians/sec ($2\pi$ rad $= 360$ degrees), and is phase-shifted by $\phi$ radians relative to an arbitrary time. Complex periodic functions, such as the pressure, flow, or diameter pulsations in a blood vessel, can be expressed mathematically as the sum of an infinite series of harmonic terms (see refs. 1, 2, 46 for details) according to the Fourier theorem:

$$F(t) = A_0 + \sum_{n=1}^{\infty} M_n \cos\left(n\omega t - \phi_n\right) \tag{28}$$

where $A_0$ is the mean value of the function (that is, the dc or zero frequency component) and $n$ is the harmonic number. The first or fundamental harmonic has frequency $\omega$ (that is, the heart rate), amplitude $M_1$ and phase lag $\phi_1$. Successive harmonics occur at integer multiples of $\omega$, each having its own amplitude $M_n$ and phase $\phi_n$. Further, each harmonic term in the series can be expressed as a combination of a sine and cosine wave of the same frequency and zero phase, using the relation:

$$M \cos\left(\omega t - \phi\right) = A \cos \omega t - B \sin \omega t \tag{29}$$

where $A = M \cos \phi$ and $B = M \sin \phi$. In haemodynamic applications pressure, flow and diameter waveforms can be adequately described by a finite series of $N$ harmonics, where $N \leqslant 10$. Equation (28) is replaced by:

$$F(t) \approx A_0 + \sum_{n=1}^{N} \left(A_n \cos n\omega t + B_n \sin n\omega t\right). \tag{30}$$

In application to the cardiovascular system it is important to remember that this theory applies to steady, periodic functions; that is, during the duration of the record being analysed, there must be a steady-state heart rate and baseline. Gessner (46) provides a discussion of this problem.

The calculation of the cosine and sine coefficients [$A_n$ and $B_n$ in equation (30)] can readily be carried out by a digital computer using a relatively simple program. The continuous signal $F(t)$ is digitized at time intervals of $(\Delta t)$. The digital series generated is $F(i\Delta t)$, for $i = 0, 1, 2, \ldots, (K - 1)$, where $K$ is the number of samples, $(K\Delta t)$ is the total time-period sampled. The maximum number of harmonic coefficients that can be obtained without aliasing errors is $K/2$, so sampling rates must be set appropriately (for additional discussion of aliasing see Chapter 3). The Fourier coefficients are then calculated according to:

$$A_0 = 1/K \sum_{i=0}^{K-1} F(i\Delta t) \tag{31}$$

$$A_n = 2/K \sum_{i=0}^{K-1} F(i\Delta t) \cos\left(2\pi ni/K\right) \tag{32}$$

$$B_n = 2/K \sum_{i=0}^{K-1} F(i\Delta t) \sin\left(2\pi ni/K\right). \tag{33}$$

The modulus ($M$) and phase ($\phi$) of each harmonic are computed from the coefficients $A$ and $B$:

$$M_n = (A_n + B_n)^{1/2} \tag{34}$$

$$\phi_n = \tan^{-1}(B_n/A_n). \tag{35}$$

If amplitude or phase corrections are necessary, because of the response characteristics of the transducers used, these can be applied to the modulus and phase of each harmonic as needed. For example, if a diameter transducer was used that introduced a phase lag of 0.1 rad/Hz, and the fundamental frequency of the heart ($f$) in the animal studied was 1.5 Hz, then the phase of each diameter harmonic would be corrected by subtracting $0.1 f_n$ from $\phi_n$.

When pressure and diameter waves are analysed, the values of $M$ at each frequency give $\Delta P$ and $\Delta R$, respectively, which can be used to calculate the dynamic incremental elastic modulus [by equation (24)] or the stress and strain increments [using equations (11) and (12)] and then $E^*$ [by equations (23) or (25)]. The difference between $\phi$ for the pressure and diameter harmonics at each frequency gives $\delta$, the phase lag of diameter relative to pressure [equations (26) and (27)].

# 4. Haemodynamics

## 4.1 Vascular elasticity and pressure wave velocity

The speed at which pressure waves travel in a fluid-filled elastic tube is governed by the distensibility of the tube wall, and the density of the fluid ($\rho$). The wave velocity in the absence of reflections ($c_0$) is often calculated by the Moens–Korteweg equations, which assumes a thin wall and no viscosity. If $h/r$ and the elastic modulus, or the pressure–volume relationship are known:

$$c_0^2 = Eh/2\rho r = (\Delta P/\Delta V)(V/\rho). \tag{36}$$

Modifications of this equation for a thick-walled tube are numerous (2) such as:

$$c_0^2 = E(R^2 - r^2)/3\rho R^2 \approx 2Eh/3\rho R \tag{37}$$

or:

$$c_0^2 = (\Delta P/\Delta R)(R/2\rho)(1 - h/R)^2. \tag{38}$$

When possible the dynamic storage modulus $E'$ should be used for $E$ in these equations. Since $\Delta V/V$, $\Delta R/R$, $h/r$ and $h/R$ are dimensionless, if $E$ or $P$ is in $N m^{-2}$ and $\rho$ is expressed in $N s^2 m^{-4}$ ($= kg m^{-3}$), then $c_0$ will be in $m sec^{-1}$. Although the effects of blood and wall viscosity are neglected, these relations will generally give good estimates of the *in vivo* pulse velocity. The pressure wavelength $\lambda$ can be calculated at each frequency $f$ from the wave equation:

$$\lambda = c_0/f. \tag{39}$$

## 4.2 Vascular impedance

### 4.2.1 Characteristic impedance

Impedance is an expression of the opposition of the arterial system to pulsatile flow. This concept quantitatively links the pressure–flow relationships in a blood vessel to its mechanical properties. For a detailed treatment and derivations see (2, 35, 46). The characteristic impedance ($Z_0$) defines the ratio of oscillatory pressure to flow in the absence of reflected waves. $Z_0$ is a function of the vessel distensibility and cross-sectional area:

$$Z_0 = \rho c_0 / \pi r^2 = (1/\pi r^2)(\rho Eh/2r)^{1/2} \tag{40}$$

using equation (36) to calculate $c_0$.

### 4.2.2 Hydraulic impedance

Reflected waves are generated in circulatory systems by non-uniformities that result in changes in $Z_0$. The major branch points in the arterial tree present the most significant mismatch in impedance, although elastic and geometric tapering of the vessels may also contribute reflections. The pressure and flow waveforms *in vivo* are thus a combination of incident and reflected waves and the $P/Q$ ratio, called the input or hydraulic impedance ($Z_x$), is quite different from $Z_0$, unless the arterial tree is similar in length or longer than the pressure wavelengths (*Figure 9*). The relation of $Z_x$ to $Z_0$ depends on the degree of reflection, attenuation and phase of the interacting waves. These factors are influenced by the viscoelasticity of the arterial wall, and the viscosity and inertia of the blood. Reflections of pressure and flow cause oscillations in $Z_x$ at multiples of the quarter-wavelength frequency $f_q$ (that is, the frequency that has wavelength $\lambda = 4L$, where $L$ is the length of the system to the main

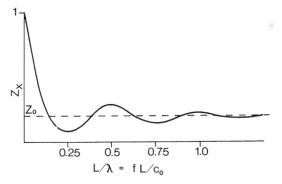

**Figure 9.** Representation of the oscillations in impedance amplitude as a function of frequency, expressed here as the system length relative to the wavelength ($L/\lambda$). The quarter wavelength frequency is at the point 0.25. The oscillations are due to reflections, and their amplitude diminishes due to viscosity. $Z_x$ approached $Z_0$ at frequencies above that where $L = \lambda$.

reflecting site). In poikilotherms, with low heart rates, $f_q$ is well above the major harmonics of the natural pulse, but in mammals $f_q$ falls between the first and second harmonic (*Figure 9*). $Z_x$ is thus a frequency-dependent parameter that can be computed from pressure and flow sinusoids as follows:

(a) Record pressure and flow waveforms simultaneously *in vivo* from the location of interest.

(b) Digitize the data, obtain pressure and flow harmonics by Fourier analysis [see equations (31–35)], and apply amplitude and phase corrections for the transducers if necessary. If the two transducers are separated by a considerable distance additional corrections may be necessary (2, 46).

(c) Calculate impedance amplitude (modulus), $Z$, and phase, $\theta$, for each harmonic frequency:

$$Z_n = P_n/Q_n, \qquad \theta_n = \beta_n - \alpha_n, \qquad (41, 42)$$

where $P_n$, $Q_n$ and $\alpha_n$, $\beta_n$ are, respectively, the amplitude and phase of the $n$th harmonic of pressure and flow.

Note that the peripheral resistance $\mathbf{R} = P_0/Q_0$, and this is the impedance at zero frequency. $Z_x$ represents the hydraulic impedance of the vasculature distal to the site of measurement. To obtain the input impedance of the whole arterial-tree pressure and flow should be measured at the outflow of the heart.

This method limits the impedance data to a small number of frequencies, the major harmonics of the $P$ and $Q$ pulses. In mammals the first 5 or 6 harmonics contain almost all of the pulsatile information. In poikilotherms this number is somewhat smaller. A larger number of frequencies in the impedance spectrum can be obtained by methods that alter the heart rate experimentally, and use spectral analysis (correlation) techniques (2, 35, 46).

### 4.2.3 Physiological significance of the impedance spectrum

The impedance spectrum characterizes the relationships of pulsatile pressure and flow in a circulatory system. It is useful for defining the haemodynamic state of an animal, for comparing different animals, for assessing the effects of vasoactive agents and experimental changes in the performance of the heart and for diagnosing circulatory disfunction.

## 4.3 Hydraulic power

The mechanical output of the heart can be measured by the PV work delivered to the blood system. The stroke work (*W*) is the product of the mean pressure and the stroke volume, or:

$$W = \int_t^T PQ\mathrm{d}t \qquad (43)$$

where $t$ and $T$ are the times of the beginning and end of the cycle. The hydraulic power is the total work done per unit time ($\dot{W}_T$), consisting of steady and oscillatory components ($\dot{W}_s$ and $\dot{W}_o$). $\dot{W}_s$ is calculated from the mean pressure and flow:

$$\dot{W}_s = P_0 Q_0. \tag{44}$$

$\dot{W}_o$ is related to the vessel elasticity through the input impedance. If pressure and flow are recorded at the outflow of the heart $\dot{W}_o$ is obtained from the Fourier components:

$$\dot{W}_o = \tfrac{1}{2} \sum_{n=1}^{N} Q_n^2 Z_n \cos \theta_n. \tag{45}$$

The oscillatory component is significant, comprising 15–20% of the total power (2). $\dot{W}_T$ also has kinetic energy components, associated with the flow velocity, but these are comparatively small.

# 5. Measurement of pressure, flow, and dimensions

An excellent description of devices to measure pressure, flow, and lengths can be found in Milnor (2). Additional details are given in the volume edited by Vincent (47).

## 5.1 Pressure

Pressure is commonly measured by an electronic transducer which has a fluid connection to the blood vessel by polyethylene or metal tubing. Typically, the transducer has a metal sensing diaphragm that is deformed by the applied pressure. This in turn deforms internal strain-gauges that are the resistors in a Wheatstone bridge circuit, and a voltage is generated in proportion to the magnitude of the pressure. The sensitivity and frequency response of this type of transducer can vary widely, making careful calibration essential before use in any experiment. As a result of developments in miniature electronics a 'disposable' type of strain-gauge pressure transducer, designed for use in hospitals (Baxter Medical Supplies), is a good choice for the applications described in this chapter. These are small, robust, inexpensive, and reusable.

Static calibration is simply done by determining the transducer output for fixed pressure heads. Dynamic calibration is not so trivial, and the user should consult the detailed analyses presented in refs. 1 and 2. A transient pulse technique, known as a 'pop' test is commonly employed as in *Protocol 5*.

---

**Protocol 5.** Transient test of a pressure transducer

**1.** Apply a transient pulse by first connecting the transducer to a small pressurized balloon, and then break the balloon cleanly by touching it with a flame.

**2.** Record the transducer response on a digital oscilloscope or high speed

**Protocol 5.** *Continued*

chart recorder. Typically, there will be a rapid drop in pressure with accompanying damped oscillations.

3. Measure the decay in amplitude of successive peak-to-peak deflections and calculate the logarithmic decrement $\Lambda$ ($\Lambda = \ln$ (peak$_x$/peak$_{x+1}$) and damped frequency $\omega_D$.

4. Calculate the damping constant $\beta_0$ from $\beta_0 = \Lambda/T$, where $T$ is the period of oscillation, the undamped frequency $\omega_0$ from $\omega_0 = (\omega_D^2 + \beta_0^2)^{1/2}$, and the damping ratio $\beta$ from $\beta = \beta_0/\omega_0$.

5. Determine the amplitude and phase response of the transducer for each frequency $\omega$, from plots of these parameters as a function of $\beta$ and the normalized frequency $\gamma$, ($\gamma = \omega/\omega_0$). These plots are available in refs. 1, 2, and 48. Alternatively, the amplitude ($A$) and phase ($\Phi$) can be calculated for a value of $\beta$ as:

$$A = 1/[(1 - \gamma^2)^2 + (2\beta\gamma)^2]^{1/2} \quad \text{and} \quad \tan\Phi = (2\beta\gamma)/(1 - \gamma^2).$$

These data can be used to determine the appropriate amplitude and phase corrections needed at each frequency. If the undamped frequency is at least ten times higher than the highest frequencies to be measured the errors become negligible.

## 5.2 Diameter and length

Diameter measurements of artery segments can be made electronically by optical, mechanical, or ultrasonic transducers. A variety of electronic calipers, using either strain-gauge beams or linear differential transformers, have been designed for use on blood vessels *in vitro* and *in vivo* (29, 49–51). The major disadvantage is that these devices must be custom-made, and also require direct contact with the vessel wall. This imposes an external load on the tissue which will limit the faithfulness of recording dynamic events.

Non-contact optical methods for diameter measurement are very useful in biomechanical studies. One type is based on a collimated light beam casting a shadow of the vessel on to a photodetector or photodiode array that is calibrated by substituting cylinders of known diameter. Details of construction of are given in (27, 52–54).

Video systems have been used to measure diameter as well as length changes on blood vessels during inflation. An analogue video dimension analyser (VDA) that is commercially available (Instrumentation for Physiology and Medicine, San Diego, California, USA), measures the distance between two contrast boundaries (e.g. the edges of the vessel or target lines on the surface) in the horizontal axis of a video image. Resolution is better than 0.1% of the horizontal field and linearity is within 0.5% (55). By combining two VDAs biaxial deformations can be recorded. The disadvantages of this instrument is high cost and relatively low frequency response. For dynamic tests and *in vivo* recording using the VDA we apply phase and amplitude corrections to each harmonic after Fourier analysis. Alternatively, the deformation event can be recorded by high-speed video and played back in slow motion for analysis by the VDA. An improved VDA system with 200 Hz response has recently been developed (56). Recent developments in computer software make the application of digital video analysis more feasible. Relatively inexpensive techniques for frame by frame image analysis of video images are widely in use, and these are well suited for dimension measurement.

The transit time of ultrasound pulses between two piezoelectric crystals (one transmitting and one receiving) fixed on the surface of a blood vessel can be used to determine diameter or length changes. This requires an instrument to generate and analyse the ultrasound signals (e.g. the sono-micrometer, Triton Technology, San Diego, California, USA), and a knowledge of the speed of sound in the transmission medium.

## 5.3 Flow

The most popular types of instruments to measure pulsatile flow are based on either electromagnetic, ultrasonic transmit-time or Doppler methods. Electromagnetic flowmeters use the Faraday principle that the movement of an electrically conductive medium (like blood) within a magnetic field generates a voltage. In practice, a cuff containing an electromagnetic coil is fit around the vessel, the coil is excited by either square or sinusoidal pulses and the induced signal is detected by electrodes in the cuff that are in direct contact with the vessel wall. Determining the zero flow output requires occlusion of the vessel, and static calibration is done by perfusion of the vessel *in vitro* with blood or saline after the *in vivo* measurements. The major disadvantage is the high cost of the system which is increased if several sizes of cuff are required. Indwelling catheter tip probes are also available.

The simplest ultrasound method involves measuring the transit time of MHz sound waves transmitted on a diagonal path through the blood vessel. Two piezoelectric crystals are used as both transmitter and receiver. They are contained in a cuff that wraps the vessel and are oriented to be on opposite sides but separated along the vessel axis. The sound transmission between these sites will be influenced by the velocity of blood, the effect adding to the transmission in one direction and subtracting from that in the opposite direction.

A more attractive ultrasound technique makes use of the Doppler effect—that is, the sound reflected from a moving body will undergo a frequency shift. In this application the blood cells act as the reflectors, and the flowmeter produces a signal proportional to the size of frequency shift, which itself is proportional to the blood velocity. The probe can be a single crystal that acts as transmitter and receiver or a pair of crystals mounted together. This technique has several attractive features: probes are inexpensive and can be used on various sized vessels, the signal is very stable, zero flow is easily determined electronically, and the probes may be used non-invasively. The disadvantage is difficulty in calibration, and the fact that the measurement is not an average of flow across the whole vessel, but only within the narrow beam produced by the crystal. A relatively inexpensive pulsed Doppler flowmeter is available commercially from the Bioengineering Group at the University of Iowa.

# References

1. McDonald, D. A. (1974). *Blood flow in arteries*. Edward Arnold, London.
2. Milnor, W. R. (1982). *Hemodynamics*. Williams & Wilkins, Baltimore.
3. Fung, Y. C. (1984). *Biodynamics: circulation*. Springer-Verlag, New York.
4. Burton, A. C. (1954). *Physiol. Rev., **34**, 619.*
5. Gordon, J. E. (1975). In *Comparative physiology: functional aspects of structural materials* (ed. L. Bolis, S. H. P. Maddrell, and K. Schmidt-Nielsen), pp. 49–57. North-Holland, Amsterdam.
6. Bogen, D. K. and McMahon, T. A. (1979). *Biophys. J.,* **27**, 301.
7. Murray, C. D. (1926). *Proc. Natl. Acad. Sci.,* **12**, 207.
8. Yin, F. C. P., Liu, Z., and Brin, K. P. (1987). In *Ventricular vascular coupling: clinical, physiological and engineering aspects* (ed. F. C. P. Yin), pp. 385–98. Springer-Verlag, New York.
9. Langille, B. L. and Jones, D. R. (1977). *J. exp. Biol.,* **68**, 1.
10. Shadwick, R. E., Gosline, J. M., and Milsom, W. K. (1987). *J. exp. Biol.,* **130**, 87.
11. Skalak, R. (1972). In *Cardiovascular fluid dynamics* (ed. D. M. Bergel), pp. 341–76. Academic Press, New York.
12. Alexander, R. McN. (1983). *Animal mechanics* (2nd edn). Blackwell Scientific Publications, Oxford.
13. Gosline, J. M., Shadwick, R. E., DeMont, M. E., and Denny, M. W. (1988). In *Biological and synthetic polymer networks* (ed. O. Kramer), pp. 57–77. Elsevier Applied Science, London.
14. Cox, R. H. (1978). *Am. J. Physiol.,* **234**, H533.
15. Humason, G. L. (1972). *Animal tissue techniques* (3rd edn). W. H. Freeman, San Francisco.
16. Horobin, R. W. and Flemming, L. (1980). *J. Microscopy,* **119**, 345.
17. Elder, H. Y. and Owen, G. (1967). *J. Zool., Lond.,* **152**, 1.
18. Hall, D. A. (ed.) (1976). *The methodology of connective tissue research*. Joynson Bruvvers.
19. Soskel, N. T., Wolt., T. B., and Sandberg, L. B. (1987). In *Methods in enzymology,* Vol. 144, (ed. L. V. Cunningham) p. 196. Academic Press, Orlando.
20. Miller, E. J. and Rhodes, R. K. (1982). In *Methods in enzymology,* Vol. 82, (ed. L. V. Cunningham and D. W. Frederiksen) p. 33. Academic Press, New York.
21. Sage, E. H. and Gray, W. R. (1979). *Comp. Biochem. Physiol. [B],* **64**, 313.
22. Berg, R. A. (1982). In *Methods in enzymology,* Vol. 82, (ed. L. V. Cunningham and D. W. Frederiksen) p. 372. Academic Press, New York.
23. Tsugita, A., Uchida, T., Mewes, H. W. and Ataka, T. (1987). *J. Biochem. (Tokyo),* **102**, 1593.

24. Shadwick, R. E. and Gosline, J. M. (1985). *J. exp. Biol.*, **114**, 239.
25. Shadwick, R. E. and Gosline, J. M. (1985). *J. exp. Biol.*, **114**, 259.
26. Shadwick, R. E., Pollock, C. M., and Stricker, S. A. (1990). *Physiol. Zool.*, **63**, 90.
27. Bergel, D. H. (1961). *J. Physiol.*, **156**, 445.
28. Bergel, D. H. (1961). *J. Physiol.*, **156**, 458.
29. Dobrin, P. B. (1970). *Circ. Res.*, **27**, 105.
30. Patel, D. J., Janicki, J. S., Vaishnov, R. N., and Young, J. T. (1973). *Circ. Res.*, **22**, 93.
31. Cox, R. H. (1974). *J. appl. Physiol.*, **36**, 381.
32. Cox, R. H. (1975). *Am. J. Physiol.*, **229**, 1371.
33. Busse, R., Bauer, R. D., Sattler, T., and Schabert, A. (1981). *Pflügers Arch.*, **390**, 113.
34. Dobrin, P. B. (1983). In *The cardiovascular system: Handbook of physiology*, pp. 65–102. Americal Physiological Society, Bethesda, Maryland.
35. Gow, B. S. (1980). In *The cardiovascular system, 2: Handbook of physiology*, pp. 353–408. American Physiological Society, Bethesda, Maryland.
36. Fung, Y. C. (1972). In *Biomechanics: its foundations and objectives* (ed. Y. C. Fung, N. Perrone, and M. Anliker), pp. 181–208. Prentice-Hall, Englewood Cliffs, NJ.
37. Ward, I. M. (1985) *Mechanical properties of solid polymers* (2nd edn). John Wiley, Chichester.
38. Attinger, F. M. L. (1968). *Circ. Res.*, **22**, 829.
39. Gosline, J. M. and Shadwick, R. E. (1982). *Pac. Sci.*, **36**, 283.
40. Shadwick, R. E. and Nilsson, E. K. (1990). *J. exp. Biol.*, **152**, 471.
41. Ferry, J. D. (1980). *Viscoelastic properties of polymers* (3rd edn). John Wiley, New York.
42. Shadwick, R. E. (1992). In *Biological materials: a practical approach* (ed. J. F. Vincent). IRL Press, Oxford.
43. Zatzman, M., Stacy, R. W., Randall, J., and Eberstein, A. (1954). *Am. J. Physiol.*, **177**, 299.
44. Wiederhielm, C. A. (1965). *Fed. Proc.*, **24**, 1075.
45. Attinger, E. O. (1969). *IEEE Trans. Biomed. Eng.*, **BME-16**, 253.
46. Gessner, U. (1972). In *Cardiovascular fluid dynamics* (ed. D. H. Bergel), pp. 315–49. Academic Press, New York.
47. Vincent, J. F. (ed.) (1992). *Biological materials: a practical approach*. IRL Press, Oxford.
48. Gabe, I. T. (1972). In *Cardiovascular fluid dynamics* (ed. D. H. Bergel), pp. 11–50. Academic Press, New York.
49. Mallos, A. J. (1962). *J. appl. Physiol.*, **17**, 131.
50. Gow, B. S. (1966). *J. appl. Physiol.*, **21**, 1122.
51. Murgo, J. P., Cox, R. H., and Peterson, L. H. (1971). *J. appl. Physiol.*, **31**, 948.
52. Wetterer, E., Busse, R., Bauer, R. D., Schabert, A., and Summa, Y. (1977). *Pflügers Arch.*, **368**, 149.
53. Schabert, A., Bauer, R. D., and Busse, R. (1980). *Pflügers Arch.*, **385**, 239.
54. Papageorgiou, G. L. and Jones, N. B. (1985). *J. Biomed. Eng.*, **7**, 295.
55. Fronek, K., Schmid-Schoenbein, G., and Fung, Y. C. (1976). *J. appl. Physiol.*, **40**, 634.
56. Geiger, D. (1988). *Rheolog. Acta*, **26**, 473.

# Hydrodynamics of animal movement

M. E. DeMONT and J. E. I. HOKKANEN

## 1. Basic introduction to fluid dynamics

All organisms, whether plant or animal, terrestrial or aquatic, sessile or freely moving, large or small, experience flow of fluid over and through external and internal structures. This flow can be an unwanted consequence of some other process that induces the flow, or generated purposefully for functional mechanisms. The forces generated by this flow, for all situations, have important implications to the life of the organism, and influence the physiology and ecology of the organism, as well as many other biological processes. Clearly, knowledge of these fluid processes will have important implications in understanding the relationship of an organism to its environment. This chapter introduces some basic theory of fluid dynamics and experimental techniques, important in understanding one particular aspect of biology—how animals move through water.

### 1.1 Definition of a fluid

Fluids are infinitely deformable and precisely follow the contours of a solid surface. The fluid layer immediately next to a solid does not slip with respect to the solid. This empirically confirmed property of fluids is called the *no-slip condition*. This can be demonstrated by fine dust adhering to a smooth table as you try to blow it away, yet it is easily swept away by the touch of a finger. The no-slip condition creates a region in the fluid where the fluid velocity increases from zero at the solid–fluid interface to the free stream velocity. This region is called the *boundary layer*. The effect of the solid–fluid interface diminishes asymptotically, in theory, extending throughout the fluid. It is customary to define the outer edge of the boundary layer to be where the fluid velocity has reached 99% of the free stream velocity.

The magnitude of the force required to deform a fluid depends on how rapidly the fluid is deformed. Viscosity is a property related to fluid deformation, that corresponds to elasticity in solids. *Dynamic viscosity* describes the internal friction of the fluid. This concept can be clarified by imagining that a fluid consists of layers of adjacent sheets. The easier the sheets slide across each other, the lower the viscosity. The units of dynamic viscosity are $kg\,m^{-1}\,sec^{-1}$. The ratio of dynamic viscosity and density appear frequently in many useful equations. This ratio is often simplified to the *kinematic viscosity*. Kinematic viscosity is expressed in the units $m^2\,sec^{-1}$. Fluids are said to be 'Newtonian' if the viscosity is independent of the magnitude of the applied stress. Water and air can be assumed to be Newtonian, but many biological fluids, such as blood, are non-Newtonian. Fluids can also be assumed to be incompressible for most biological processes.

## 1.2 Properties of water

The physical properties of water are temperature-dependent. Both density and dynamic viscosity decrease as temperature increases. Representative values for dynamic and kinematic viscosity are given in *Table 1* for fresh water and sea water.

**Table 1.** Physical properties of fresh and sea water (at a salinity of 34.8%) (Data from ref. 3)

|  | Temperature (°C) | Dynamic viscosity ($kg\,m^{-1}\,sec^{-1}$) | Kinematic viscosity ($m^2\,sec^{-1}$) |
|---|---|---|---|
| Fresh water | 10 | $1.307 \times 10^{-3}$ | $1.308 \times 10^{-6}$ |
|  | 20 | $1.002 \times 10^{-3}$ | $1.004 \times 10^{-6}$ |
|  | 30 | $0.798 \times 10^{-3}$ | $0.801 \times 10^{-6}$ |
| Sea water | 10 | $1.391 \times 10^{-3}$ | $1.358 \times 10^{-6}$ |
|  | 20 | $1.072 \times 10^{-3}$ | $1.047 \times 10^{-6}$ |
|  | 30 | $0.868 \times 10^{-3}$ | $0.848 \times 10^{-6}$ |

## 1.3 Hydrodynamic forces

### 1.3.1 Friction drag and pressure drag

*Ideal fluids* have zero viscosity, and thus no internal friction. Real fluids, however, always have some viscosity. Thus, flow past a solid surface will always exert a force on the solid, due to the internal friction generated with the stationary layer next to the solid. The drag force arising from this inter-lamellar friction of fluids is called the *friction drag*.

Let us consider an ideal fluid flowing past a cylinder. The axis of the cylinder is assumed to be perpendicular to the flow (*Figure 1*). The flow is undisturbed far from the object, both upstream and downstream. Fluid particles moving in a path that is intercepted by the cylinder will slow down as they approach the object, since it blocks the mainstream flow. The particles have to speed up as they go around the object, in order to catch up with the rest of the flow. At the rear, the fluid particles slow down again before leaving the

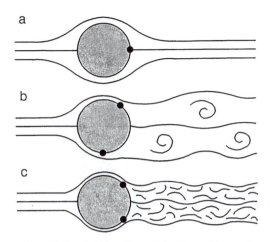

**Figure 1.** Flow around a cylinder. (a) The value of the Reynolds number for the flow is less than 10. This can also represent flow in an ideal fluid for any Reynolds number. (b) *Re* is about 1000, and in (c) *Re* is above 200 000. The bullet markers show the separation points of the fluid on the cylinder surface. (Redrawn from ref. 3, with permission of Wadsworth, Inc.)

vicinity of the object. The flow pattern at the rear is the same as the flow pattern at the front (*Figure 1a*). The total energy in this system is conserved, therefore a reduction in fluid speed causes a decrease in kinetic energy, with a concomitant increase in potential energy. If the flow is horizontal, the increase in the potential energy is seen as an increase in pressure. Thus, the low flow speeds seen at the front and rear of the cylinder imply high pressures at those locations. Since the flow patterns are similar, the forces generated cancel each other out, so that no net force acts on the cylinder in an ideal fluid.

Real fluid loses energy in the form of heat on its way around the cylinder, due to frictional shear forces generated in the boundary layer (*Figure 1b, c*). The fluid travels towards a region of increasing pressure after passing halfway round the cylinder. The fluid does not have enough kinetic energy to make it all the way, since some of the kinetic energy is lost as heat. It is eventually brought to rest, and from there on the flow *separates* from the surface of the cylinder. Beyond the separation point, eddies are formed and the pressure does not continue to rise, remaining lower than the pressure at the upstream face of the cylinder. This pressure difference creates a net force directed in the downstream direction, and is called the *pressure drag*.

The total drag force ($F_D$) is usually written in the following form, which defines the drag coefficient:

$$F_D = \tfrac{1}{2}\rho S C_D U^2 \tag{1}$$

where $\rho$ is the density of the fluid, $S$ is a surface area (see below), $C_D$ is the drag coefficient, and $U$ is the velocity. This deceptively simple relationship is complicated since the drag coefficient is not a constant, but depends on some of the parameters in the equation. These relationships are discussed in Section 1.5. Several methods are used to define the surface area, and are discussed in Section 8.

### 1.3.2 Added mass and acceleration reaction

If an organism is accelerating in a stationary fluid, an extra force associated with the acceleration occurs. This force, called the *acceleration reaction*, is dependent on the change in velocity of the body relative to the fluid. The acceleration reaction also exists in ideal fluid, where friction and pressure drag are absent. The acceleration reaction is the force required to accelerate the fluid as it moves past the organism. It depends on the size, shape, and acceleration of the body, and is written as:

$$F_A = \rho V C_A a \tag{2}$$

where $\rho$ is the density of the fluid, $V$ is the volume of the body (i.e. the volume of the fluid the body displaces), $C_A$ is the dimensionless *added mass coefficient*, and $a$ is the acceleration. The mass of fluid ($C_A \rho V$) accelerated simultaneously with the body is called the *added mass*. Added mass coefficients depend on the shape of the body, and on the overall pattern of flow (see Section 1.5).

Denny (1) describes the situation for sessile organisms in which the body is stationary but the flow accelerates past it. In this case, the acceleration reaction is present, but an additional force, $\rho V a$, emerges since the body occupies a volume $V$ of the fluid which is not accelerating with the rest of the fluid. This additional inertial force is called the *virtual buoyancy*. In this situation, the acceleration reaction becomes:

$$F_A = \rho V C_M a \tag{3}$$

where $C_M$ is the dimensionless *inertia coefficient* (note that $C_M = C_A + 1$). The acceleration reaction resists both acceleration and deceleration, in contrast to drag which resists acceleration but augments deceleration.

### 1.3.3 Lift

As noted earlier, a fluid speeds up as it passes around an object blocking the mainstream flow. If the object is not symmetrical, the flow across the slide with the longer surface contour experiences higher speeds of flow. This creates an asymmetric distribution of pressure, lowest on the side with the longest contour, so that the forces exerted on the object do not cancel. A net force is produced that is directed towards the region with the longest contour. This force is perpendicular to the direction of flow, and is called *lift*.

The total lift force ($F_L$) is written in the following form, which defines the lift coefficient:

$$F_L = \tfrac{1}{2}\rho S C_L U^2 \tag{4}$$

where $S$ is the area projected perpendicularly to the flow, and $C_L$ is the lift coefficient.

## 1.4 Reynolds number

Inertial forces tend to keep moving objects in motion, while viscous forces oppose the persistence of motion. The physical character of the flow of a fluid is determined by the relative importance of these two forces. *Reynolds number* (*Re*) is the ratio of inertial and viscous forces, and is formally defined as:

$$Re = \frac{\rho l U}{\mu} \tag{5}$$

where $\rho$ is the density of the fluid, $l$ is a linear dimension characteristic to the situation, $U$ is the velocity of flow, and $\mu$ is the dynamic viscosity of the fluid. For open flow situations, the longest length of the object in the direction of flow is used as the characteristic linear dimension. The diameter of the pipe is used for flow through a pipe. It is useful to note that equality of Reynolds number for flows passing similarly shaped objects guarantees that the patterns of flow will be the same. This idea is discussed in more detail in Section 4.

An important aspect of the Reynolds number is that it indicates when fluid flow changes from *laminar* to *turbulent*. As long as the flow is laminar, fluid particles move in a smooth, orderly fashion and any disturbances are damped out by viscosity. Fluid motion is similar on both small and large scales at low Reynolds numbers. As the Reynolds number increases, the viscous damping cannot keep pace with the developing disturbances, so eddies and other chaotic discontinuities develop and grow. The flow changes to a turbulent system, where the large-scale movement is in the direction of flow, but small-scale fluctuations of velocity in directions other than the main flow exist. Flow through a pipe, for example, changes from a laminar to a turbulent regime at Reynolds numbers of about 2000. The flow can be kept laminar at somewhat higher Reynolds numbers if extra precautions are taken not to provoke any disturbances. Reynolds numbers are useful only for qualitative comparisons, so there is little reason to express them to more than two significant figures. Reynolds numbers for swimming organisms cover an enormous range, varying from 300 000 000 for a blue whale to 0.03 for a sperm, and down to 0.00001 and below for an individual cilia or flagella. For most fish, the Reynolds number is between 10 000 and 10 000 000.

## 1.5 Force coefficients and Reynolds number

### 1.5.1 Drag coefficients

The drag coefficient is not a constant, but depends on several factors, including the size of the object and the velocity of flow. This relationship is complex, even with an object as simple as a cylinder oriented perpendicular to the flow. Many of the peculiarities can be lumped into one relationship by examining how the drag coefficient varies with the Reynolds number. Our previous discussion about the origin of pressure drag around a cylinder can help to understand this phenomenon. It is obvious that both inertia of the fluid and viscous friction affect the location of the separation point of the fluid flowing past the cylinder. The pressure at the rear of the object diminishes as the separation point on the surface creeps upstream during increasing flow velocity, showing that the drag coefficient and the Reynolds number are related. The complexities of the relationship can be seen by examining in more detail the processes that occur around the cylinder during crucial transitions in the flow regimes.

Friction drag dominates at Reynolds numbers of less than about 1, and its effects are noticeable over long distances. Measurements of the drag force on the cylinder may give exaggerated results if there is a wall even a large number of cylinder-diameters away, because of the enhanced friction induced by the overlapping boundary layers. The flow around the cylinder is laminar as long as the Reynolds number stays below 10 (*Figure 1a*), above that, eddies start to form (*Figure 1b*). The drag coefficient decreases with increasing Reynolds number, but not fast enough to prevent the drag force from increasing. The boundary layer around the cylinder remains laminar until a critical Reynolds number around 100 000 is reached, where it then turns turbulent (*Figure 1c*). Turbulent boundary layers are less susceptible to separation, so the separation point moves downstream and the drag coefficient drops as the pressure drag decreases. For smooth, uniform surfaces this transition can be sudden and dramatic, resulting in a decrease in the total drag. The change takes place more gradually for rougher surfaces usually encountered in nature. For a cylinder, pressure drag makes up 57% of the total drag at $Re = 10$, 71% at $Re = 100$, 87% at $Re = 1000$, and 97% at $Re = 10 000$.

Some empirical equations have been derived for the relationship of drag coefficient and Reynolds number for simple, uniform structures. For example, for a rigid plate oriented parallel to the flow, and thus with negligible pressure drag, the relationship between the Reynolds number and the friction drag coefficient has been estimated to be:

$$C_D = 1.33 Re^{-0.5} \tag{6}$$

when the flow is laminar, and

$$C_D = 0.072 Re^{-0.2} \tag{7}$$

when turbulent (2). Flexible plates passively flapping in the flow [for example, kelp fronds (1)] have higher $C_D$ values.

Boundary layer thickness decreases with increasing Reynolds number. For a flat plate or a slightly curved surface, assuming that no separation of flow occurs, the thickness of the boundary layer ($\delta$) at a distance $l$ downstream from the leading edge of the plate is given by:

$$\delta = 5 \sqrt{\frac{1}{Re_1}} \tag{8}$$

where $Re_l$ is the local Reynolds number, calculated using $l$ as the characteristic length. The relation holds for Reynolds number values encountered by fish. For more discussion, see Vogel (3).

## 1.5.2 Added mass and inertia coefficients

Far too little work has been done yet to determine added mass or inertia coefficients for biological shapes, considering their importance for analysis of swimming (see Section 10). Daniel (4) experimentally measured added mass coefficients for ellipsoid and hemiellipsoid (jellyfish-shaped) models for studies on unsteady medusan swimming. The added mass coefficients were dependent on the fineness ratio of the model, and varied from 0.2 to 2.5 for fineness ratios between 3.0 to 0.5. The fineness ratio was defined as the ratio of the length of the axis of revolution to the diameter (ellipsoidal models) or to the radius (hemiellipsoids). Denny (1) lists theoretical values of added mass coefficients for spheres, cylinders, and plates. The added mass coefficient of a long smooth cylinder far away from any solid surface is 1, decreasing to 0.6 for a short cylinder of length-to-diameter of 1.2.

Inertia coefficients range from 1.3 to 1.8 for limpets, snails, and barnacles in flow, which operate at $Re$ of about 100 000 (1). Roughness of the animal's surface can affect the value of the inertia coefficient. For a smooth cylinder, the inertia coefficient increases from one at $Re$ of 30 000 to a maximum value (about 1.9) at $Re$ of 1 000 000, and plateauing at $Re$ above this. The rougher the surface of the cylinder, the faster the inertia coefficient reaches its peak value (between 1.9 and 2.0). As with the smooth cylinder, the inertia coefficient drops only slightly at higher values of $Re$ (1).

## 1.5.3 Lift coefficient

Lift forces can be generated by the pectoral fins of negatively buoyant aquatic species to overcome gravity. Lift forces can also be used as a propulsive force; for example, thrust is generated by the caudal fin of tuna with a lift based mechanism (5). The lift coefficient for hydrofoils depends on the angle of attack ($\alpha$) and the Reynolds number. For a given $Re$, the lift coefficient increases as $\alpha$ increases, up to a maximum of about 20 degrees. At greater angles of attack, the flow separates from the upper surface of the hydrofoil, and the lift drops abruptly—the hydrofoil stalls. The greatest lift coefficient for ordinary hydrofoils is about 1.5 at $Re$ of 1 000 000 and 1.0 for $Re$ around 1000. More information can be found in Alexander and Goldspink (5). See Chapter 5 for a discussion of aerodynamic lift and wing theory in relation to animal flight.

For limpets, $C_L$ is around 0.4 and varies little over the range of $Re$ from 1000 to 10 000. The $C_L$ is between 0.15 and 0.25 for cones of height-to-diameter ratios of between 0.9 and 0.3, respectively, over the same range of $Re$ (1).

## 2. Size and shape

Reynolds numbers typical of large animals are much larger than those for smaller organisms due to *Re* being proportional not only to the length of an animal, but also to its velocity. Since small animals swim at much lower absolute velocities than those of large animals, they operate at Reynolds numbers that are many orders of magnitude smaller. Thus the size of an organism can radically alter the type of flow forces that it experiences. This has lead to the evolution of a variety of shapes designed to accommodate particular aspects of fluid flow over an animal's body surface.

An example might help to clarify this statement. Actively swimming fish are streamlined to reduce the total drag force on their bodies. Streamlining implies that the body trunk follows the trajectories of fluid particles passing them. The point of separation for water passing over these fish is pushed as far back to the rear as possible. This minimizes the pressure drag. A mackerel, for example, has a drag coefficient close to that of a flat plate parallel to the flow.

At the high Reynolds numbers experienced by actively swimming fish, such as a mackerel, the dominant form of drag would be pressure drag. Hence it is advantageous for these animals to be streamlined. Friction drag still exists, and will in fact be increased by streamlining, since the total surface area is increased. For round bodies at high Reynolds numbers, friction drag is considerably smaller than pressure drag, so the total drag force will be reduced by streamlining. When Reynolds number approaches unity (corresponding to very small animals and low swimming velocities), friction drag begins to dominate over the pressure drag for all shapes, and streamlining is less profitable as a means of reducing total drag. Consequently, round body shapes with relatively little surface area are preferred, and are typical of very small organisms, operating at $Re < 1$.

## 3. Dimensional analysis and dimensionless products

*Dimensional analysis* can provide a partial solution to nearly any physical problem (6). The analysis provides a set of *dimensionless products* that reduce the number of variables in the problem, but the complete solution usually requires experimentation to obtain the remaining parameters and functions. The cornerstone of dimensional analysis is Buckingham's theorem, which states that equations can be reduced to a simpler relationship using dimensionless products.

To demonstrate Buckingham's theorem and the mathematical technique used, terminal velocity of a body sinking in a fluid is derived. To begin, all relevant variables involved with the problem have to be identified. In dimensional analysis, an understanding about the physics behind the particular phenomenon in question is important. Our intuition tells us that the terminal velocity ($U$) will depend on the radius of the body ($r$), the density ($\rho$) and the viscosity ($\mu$) of the fluid, and the acceleration due to gravity ($g$). Buckingham's theorem says that the relationship $f(U, r, \rho, \mu, g) = 0$ can be simplified to a set of independent products of the original parameters.

The commonly used symbol for the *dimension* of mass is $M$, for length $L$, and for time $T$. (The corresponding *units* in SI system are kg, m, and sec.) The dimension of velocity is thus $LT^{-1}$, of acceleration $LT^{-2}$, of density $ML^{-3}$, and of viscosity $ML^{-1}T^{-1}$. Any product of the variables involved must be of the form $U^a r^b g^c \rho^d \mu^e$, and correspondingly of the dimension $(LT^{-1})^a(L)^b$ $(LT^{-2})^c(ML^{-3})^d(ML^{-1}T^{-1})^e$. In order for the product to be dimensionless, the sum of the exponents of $M$, $L$, and $T$ must each be zero:

$$d + e = 0$$
$$a + b + c - 3d - e = 0$$
$$- a \quad - 2c \quad \quad - e = 0$$

In terms of $a$ and $d$, the solution becomes $b = (3/2)d - (1/2)a$, $c = (1/2)d - (1/2)a$, $e = - d$. One dimensionless product is obtained by setting $a = 1$, $d = 0$. Another dimensionless product is obtained when $a = 0$, $d = 1$. The products are $U/(r\,g)^{1/2}$ and $r^{3/2}g^{1/2}\rho/\mu$, respectively, and are independent, since they include different variables. According to Buckingham's theorem, the problem may now be written in a form $f(U/(r\,g)^{1/2}, r^{3/2}g^{1/2}\rho/\mu) = 0$. Assuming that we can solve the first variable in terms of the other, the equation for the terminal velocity becomes $U = (r\,g)^{1/2}f(r^{3/2}g^{1/2}\rho/\mu)$. The remaining function can be solved experimentally by plotting the first dimensionless product against the other.

## 3.1 Dimensionless numbers frequently useful in fluid mechanics applicable to biological systems

The variables frequently encountered in biological fluid problems are force $F$, length $l$, velocity $U$, acceleration $a$, density $\rho$, dynamic viscosity $\mu$, acceleration of gravity $g$, and surface tension $\sigma$. Five independent dimensionless products can be formed from these eight variables: Reynolds number $Re = \rho l U/\mu$, pressure coefficient $C_P = F/(\rho l^2 U^2)$, added mass coefficient $C_A = 2F/(\rho l^3 a)$, Froude number $Fr = U^2/(gl)$, and Weber number $We = \rho l U^2/\sigma$. This set is the commonly adopted one, but is not the only possible set of five independent products. In the previous example, $Fr$ and $Re$ would have emerged if we had chosen $a = 2$, $d = 0$, and $a = 1$, $d = 1$, respectively. The reason for choosing different values was to look for dimensionless products with the dependent variable $U$ in only one of them, thus providing a solution for $U$ in terms of the rest of the variables.

The five independent dimensionless products are each important for different situations. If waves and surface tension on a free water surface are not important, then Froude and Weber numbers are usually not needed, as is usually the case for most swimming animals (except surface swimmers such as ducks). When the flow relative to the object is oscillatory, two more independent dimensionless products can readily be constructed. If $T$ is the period and $f$ is the frequency of the oscillation, then the period parameter ($K = U\,T/L$) and Strouhal number ($St = f\,L/U$) (sometimes called reduced frequency parameter) should be considered. When $K$ exceeds 30, the flow pattern approaches that of a steady unidirectional flow. It was seen in Section 1.3 that forces due to different pressures require coefficients of their own ($C_D$, $C_L$).

# 4. Use of models

Physical models that can be placed in wind tunnels or flow tanks can be useful to biologists interested in examining flow phenomena affecting complex biological shapes. Scaled-up or scaled-down models of almost any biological system, from plankton to whales, can be used to examine flow patterns around structures that would normally be difficult to manipulate experimentally because of their size. The models can be constructed of almost any suitable material; machined aluminium blocks have modelled squid (7), cellulose acetate sheets frog's feet (8), and M&M candies plankton (3). The flow phenomena experienced by the model and the real animal will be the same if the values of the independent dimensionless products are the same for the model and the animal. As mentioned in the previous section, for flows not affected by surface phenomena (and for velocities well below the speed of sound) the crucial parameter is the Reynolds number. Thus, a biological system and its model will have similar flow patterns if the two have the same shape (*geometrically similar*) and the same Reynolds number. The fluid in which the model is exposed does not have to be the same as that which the original organism experiences; indeed, it is often useful to change the medium to facilitate flow visualization or achieve reasonable velocities of flow. If the fluid is changed the drag forces on the original system and model will be different; however, the drag coefficients will remain the same. Vogel (3) lists factors by which the drag force on an object is increased if it is shifted from air to a liquid. The use of physical models to examine flow phenomena can be a powerful technique, that is remarkably under-utilized by biologists.

Geometrically similar objects have the same shape, thus the ratio of corresponding lengths are the same (see also Chapter 2). For example, if the length and thickness of a fish are $l$ and $d$, respectively, then a model of the fish will be geometrically similar if:

$$\frac{l}{l_m} = \frac{d}{d_m} \tag{9}$$

272

where $l_m$ and $d_m$ are corresponding lengths of the model. Geometric similarity also implies that corresponding areas $(S)$ will vary with the square of their linear dimension, and that volumes $(V)$ vary with the cubes of their linear dimensions; thus, for example:

$$\frac{V}{V_m} = \left(\frac{d}{d_m}\right)^3 = \left(\frac{l}{l_m}\right)^3. \tag{10}$$

## 5. Flow tanks

*Flow tanks* have a variety of uses for anyone interested in applying fluid mechanics to biology. Force transducers can be used to measure drag on objects suspended in flow tanks. Film or video analysis of organisms confined to flow tanks can provide useful information on such questions as orientation and behaviour to flow of sessile organisms, or hydrodynamics of swimming in fish. Simple *flow visualization* techniques allow observation of flow patterns around submerged organisms.

Flow tanks in many engineering laboratories function by flow of water from two tanks that differ in elevation, with water being pumped back to the higher tank. The working area of such a system is a trough between the two tanks. This system, although capable of producing flow with great uniformity, requires a rather large pump (and power consumption) to maintain reasonably moderate flow velocities.

Vogel and LaBarbera (9) describe in some detail the design of a simple, but versatile flow tank (see *Figure 2*) that has proven useful for many applications in biology. The design of this flow tank uses a closed, recirculating system, thus eliminating the downstream waterfall. The best design uses an arrangement in which the water returning from the downstream end to the upstream end of the working section passes through a conduit located beneath the working section. Circulation is induced by a slowly revolving axial propeller.

The details of construction are dependent on the particular requirements of use; however, some general rules are useful to apply when constructing the tank. The working section should have a length-to-width ratio of about 10, while the depth of water should be about equal to the width of the section. An array of parallel pipes should be inserted upstream of the working section to minimize the circular component of flow induced by the rotating propeller. The propeller should be located at the downstream end of the working section, and in the descending arm of the return conduit. This position will minimize fan-induced turbulence in the working section, and alleviate problems of sealing shafts.

## 6. Visualization of flow

Flow visualization can be a useful technique to give both qualitative information about flows around irregular-shaped organisms, and quantitative data

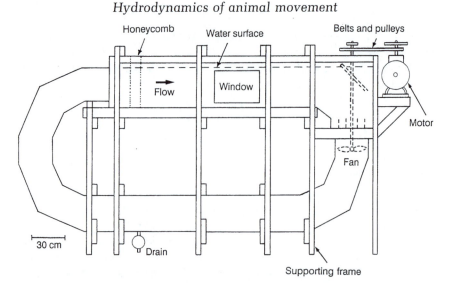

**Figure 2.** A simple flow tank that uses a closed, recirculating system. The water is circulated by an externally driven propeller, that moves the water in a clockwise direction. An array of parallel pipes (honeycomb) is placed at the upstream end of the working section, which is at the top of the tank. (From ref. 3, with permission of Wadsworth, Inc.)

about speed and direction of flow. Two techniques are used to visualize flow, both of which are most commonly used in flow tanks. *Pathlines* can be created by photographing with time or multiple exposures a visible marker that is released upstream of the flow under investigation, while *streaklines* are formed by instantaneous photographs of a continuous stream of markers that have been released upstream. In steady flows pathlines and streaklines both mark *streamlines*. Streamlines are curves drawn tangent to any point in the direction of the velocity vector at that point. These can be very useful to biologists studying flow around an organism, since streamlines obey the principle of continuity. Thus, when two streamlines diverge, the velocity of flow has decreased. In unsteady flows, neither pathlines or streaklines mark streamlines.

Three different techniques are commonly used to produce markers. The simplest approach is to use a hypodermic syringe to inject a dye into the upstream flow. Continuous release of the dye will mark a streakline. The rate of injection should be low enough so that the free stream flow is not affected, yet high enough to counter the wake of the syringe tip. The choice of dye can depend on several factors, such as the toxicity of the dye to the organism, the affinity of the dye to the material from which the flow tank is constructed, and of course the cost. Rhodamine B and milk are two commonly used dyes. The second approach is to release small, neutrally buoyant particles in the upstream flow. A time-exposed photograph will mark a pathline, while the distance between points on photographs taken with repetitive stroboscopes

**274**

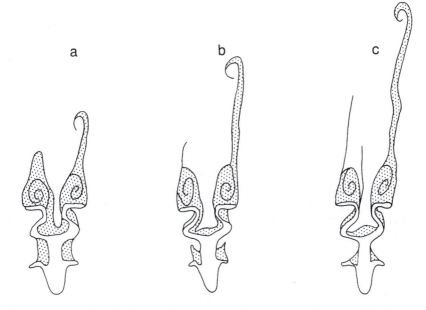

**Figure 3.** A flow visualization experiment using a frog specimen (*Hymenochirus boettgeri*). The stippled regions indicate regions of coloured glycerol released as the animal dropped through a tank of clear glycerol. The time between each illustration is 0.5 sec. (From ref. 8, with permission of the National Research Council of Canada.)

are measures of the speed of the local flow. The particles can be manmade, such as latex spheres, or natural, such as unialgal cultures. The last technique is to produce hydrogen bubbles by inserting a fine wire as a negative electrode of a DC electric circuit in the upstream flow. This technique is flexible, since short insulated sections of wire can be used to create space markers, and pulsating flow to produce time markers. Some limitations exist, however, since the bubbles tend to redissolve, and also rise slowly.

Flow visualization can also be useful in drop tank experiments. Test specimens are soaked in a 40% glycerol solution, that is stained with fuchsin. The specimen is then dropped in a tank of clear glycerol, producing visible streams of coloured glycerol, that can be filmed for later analysis. *Figure 3* shows an example of a flow visualization experiment with a frog specimen.

# 7. Fluid measurements

## 7.1 Measurement of fluid properties

The fluid properties most commonly required for biological research are density and viscosity. On most occasions, it is not necessary to measure these properties, but simply look up the appropriate value in a table (10). The simplest technique to measure viscosity is to use a device called an Ostwald viscometer. Details of these, plus other techniques, are given in Vennard and Street (11).

## 7.2 Measurement of net flow

Net flow is simply the volume of water passing a given point in a given period of time. For contained steady flow, the flow rate can easily be measured by collecting the total quantity of fluid in a measured time. The collections can be made by weight or volume, but this means of measuring flow has limited applications in situations other than special laboratory circumstances with small flows.

A *Venturi meter* can be used as an inexpensive in-line volume flow meter. An example is shown in *Figure 4*. It contains a smooth entrance cone, a short cylindrical section, and a downstream diffuser. The pressures at the entrance to the meter (at 1), and in the throat (at 2), can be measured with pressure transducers or manometers. The following equation can be used to derive the velocity at the entrance to the meter ($U_1$):

$$\Delta p = \frac{\rho U_1^2}{2}\left(1 - \frac{S_1^2}{S_2^2}\right)$$  (11)

where $\Delta p$ is the difference in pressure, and $S_1$ and $S_2$ are the cross-sectional areas of the tube where pressure is measured. The volume flow rate is then simply $U_1 S_1$ (m$^3$ sec$^{-1}$).

Dilution methods can be used for field measurements of flow rate in small open streams. For example, a concentrated salt solution ($C_1$) is added at a rate $Q_s$, to a flow with an unknown rate ($Q$). The concentration of salt in the unknown flow ($C_2$) is measured downstream after complete mixing. If $Q_s$ is small compared to the unknown flow, then $Q$ is given by:

$$Q = Q_s\left(\frac{C_1}{C_2}\right)$$  (12)

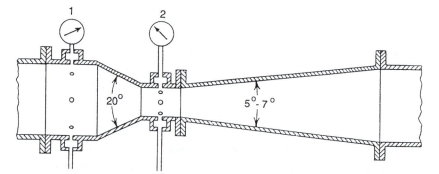

**Figure 4.** A cross-section of a Venturi meter. (From ref. 11, reprinted with permission of John Wiley and Sons, Inc. All Rights Reserved.)

The average mainstream flow in an open channel can be estimated with a simple, inexpensive technique using moulded plaster of Paris casts placed in the environment. The casts are weighted before placement, recovered one or two days later, and then dried and reweighed. The loss in weight is a measure of the net flux. The casts are calibrated by placing identical samples in known water velocities. Some practical problems exist, however, since the cast has to be radially symmetric, and the rate of dissolution is temperature-dependent.

## 7.3 Measurement of flow velocity

Measurements of the velocity of flow are extremely important for many aspects of fluid problems. There are several inexpensive and reliable methods to make these measurements, in both field and laboratory conditions. However, the techniques are generally not the same, and some care must be taken to ensure accurate measurements in either situation. Two types of velocity measurements can be made: measurement of maximum water velocity, and continuous water velocity measurement. The first is important for organisms exposed to wave-swept environments, and these techniques have been described in detail in Denny (1).

The simplest technique involves a modified spring scale, to which a thin disk is attached via a stiff cable connected at the centre of the disk. The whole apparatus is attached to the substrate in exposed environments. The scale records the maximum force exerted on the disk by flow over the apparatus. The maximum velocity can be calculated with knowledge of the drag coefficient of the disk, which is about 1.2, and independent of the Reynolds number of typical wave-swept environments. The response time of this device is not high, since reorientation requires some time.

*Figure 5* shows a device with a faster response time, which can also record

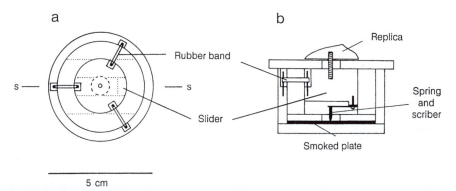

**Figure 5.** A simple device for measuring the maximum force exerted on intertidal organisms. (a) A top view showing the central movable slider supported by three rubber bands. (b) A side view through section s – s showing the model of the organism attached to the slider. The movable slider scribes the smoked plate as forces are imposed on the replica. (From ref. 23, with permission of the American Society for Limnology and Oceanography, Inc.)

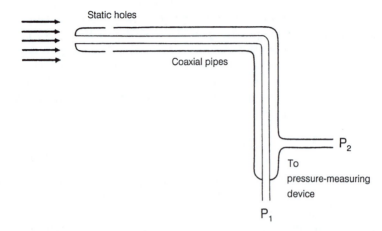

**Figure 6.** A schematic diagram of a Pitot-static tube, showing a longitudinal section. One aperture faces the upstream flow while smaller holes are placed parallel to the flow. (From ref. 3, with permission of Wadsworth, Inc.)

the direction that the maximum force was exerted. This device consists of a Teflon slider that is suspended by rubber bands. A replica of an organism is attached to the slider, and exposed to the open flow. Flow forces applied to the replica move the slider, which scribes a smoked plate secure beneath it. Calibration of the device is easily performed in the laboratory by hanging known weights from the slider while the device is placed sideways. If the drag coefficient of the attached replica is known, then the maximum velocities can be calculated from the maximum force. The response time of this device increases in proportion to the ratio of the mass of the slider and attached replica, and the stiffness of the rubber bands.

Laboratory measurements of continuous water velocity can be made with the classic *Pitot-static tube*. *Figure 6* shows a schematic diagram of such a tube. The principle of operation is based on a comparison of dynamic and static pressures, each of which is measured at a different aperture. The aperture facing upstream measures the total pressure ($P_1$), which represents the sum of the static and dynamic pressure. The other hole measures only the static pressure ($P_2$). The pressure difference between the two points ($\Delta p$) can be used to calculate the velocity as follows:

$$\Delta p = \frac{\rho U^2}{2} \tag{13}$$

where $\rho$ is the density of the fluid. For use with an attached manometer the following practical equation can be used to calculate the velocity:

$$U^2 = \frac{2g\Delta h(\rho_m - \rho_a)}{\rho_a} \tag{14}$$

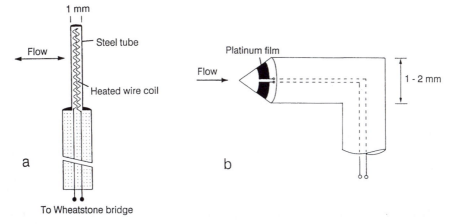

**Figure 7.** A schematic diagram of (a) a metal-clad heated probe, and (b) a hot-film flow probe. (From ref. 1, with permission of Princeton University Press.)

where $g$ is the gravitational constant, $\Delta h$ is the difference in height of the liquid in the manometer, and $\rho_m$ and $\rho_a$ are the densities of the manometric and ambient fluids, respectively.

There are several other techniques available to make continuous measurements of flow velocity. The suitability of a specific technique depends on a number of factors, including the spatial and temporal scale of the velocity measurements required. Propeller flow meters, that are based on the principle that the angular velocity of the propeller is proportional to the velocity of flow, are commercially available (Omega Engineering Inc., Stamford, Connecticut). These meters tend to average the velocity over the diameter of the propeller, so if small-scale measurements are required, meters with small propellers should be used. The response time is dependent on the rotational angular momentum of the propeller. Electromagnetic flow meters are available that are useful for measuring flow velocity through pipes. Force transducers, such as those described in Section 8, can be used to produce continuous records of force, which in turn can be converted to velocity measurements if suitable drag coefficients for the attached model are known. Heated probe flow meters are also available, and work by heating an element that is situated in the flow being measured. The faster the flow over the heated element, the greater the heat loss to the environment. With suitable electronic circuits, the heat loss can be converted to an electrical signal that can be continuously monitored. Several variations of these probes exist, and include hot wire, metal-clad hot wire, hot film, and thermistor. Two examples are shown in *Figure 7*. A thermistor flow meter designed by LaBarbera and Vogel (12) has proved useful for measuring low velocity flows. This model has the advantage of being inexpensive, rugged, and sensitive to low speeds. Vogel (3) gives a

**279**

correction to the original circuit design. Denny (1) lists commercial suppliers for many of these meters.

# 8. Measurement of drag and lift forces

Drop tank experiments provide an inexpensive technique to measure drag forces and drag coefficients. The potential of this technique is demonstrated by the exemplary work of Gal and Blake (8). Specimens are generally preserved in functional orientations, before free falling in a vertical tank of water. At terminal velocity, the submerged weight of the specimen $W_s$, is equal to the drag force, $F_D$:

$$W_s = F_D = \tfrac{1}{2}\rho S U^2 C_D \qquad (15)$$

where $\rho$, $S$, $U$, and $C_D$ were defined previously (see Section 1.3.1). The submerged weight is calculated from:

$$W_s = W_a \left(1 - \frac{\rho}{\rho_s}\right) \qquad (16)$$

where $W_a$ is the weight of the specimen in air, and $\rho_s$ is the density of the specimen. Lead weights can be inserted into the body cavity to change the submerged weight. This process will produce a range of terminal velocities, which can be calculated from frame by frame analysis of digitized time-displacement data. Equation (15) can then be used to calculate the drag force and drag coefficients. Two reference areas commonly used are the total surface or 'wetted' area and maximum area projected on to a plane normal to the direction of flow ('frontal area'). The wetted area can be estimated by approximating specimens as a series of cones, cylinders, or spheroids, which have mensuration formulas for surface area (13). Frontal area can be calculated from digitized photographic or video images. Two other reference areas are also used: the maximum projected area for airfoils ('plan' or 'profile'), and two-thirds power of volume, which may be the most biologically appropriate. It is important to state the reference area used when calculating the drag coefficient, since the choice for the area in equation (15) can have significant effects on the calculated value of the drag coefficient (14). For example, the drag coefficient for a streamlined fish will be more than ten times larger if the frontal area is used instead of the wetted area.

The drag force on large organisms can be obtained by attaching frozen specimens to a harness and towing the entire set-up in water channels. The set-up is attached via a stiff mechanical arm to an electronic load cell, which can be calibrated by suspending weights of known mass. Both subsurface and surface drag forces can be measured if the setup is towed at various depths. The depth should be more than about three body diameters for subsurface measurements that are free of surface effects. The drag of the harness should be subtracted from the measured drag.

Drag forces can also be measured by attaching rigid organisms (or models of organisms) to submersible transducers. The transducer contains foil strain-gauges that are oriented to deform as the organism experiences drag (see Chapter 3 for more information on the use of strain-gauges). The whole set-up is placed in the working section of a flow tank, where it can be subjected to different velocities of flow. Lift forces can be measured using this technique, but in this situation the gauges are designed to deform to forces of lift, that are oriented perpendicular to the direction of flow. *Figure 8* shows examples of both of these transducers.

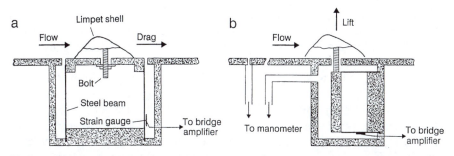

**Figure 8.** Electronic transducers that measure lift and drag forces on intertidal animals. (a) This transducer measures drag. The force applied to the shell deforms the platform that is supported by steel beams. Foil strain-gauges are attached to the beams. These transduce the deformation of the steel beams into a proportional voltage. (b) This transducer measures lift by arranging the platform to deform when forces perpendicular to the flow are applied. (From ref. 24, with permission of the National Research Council of Canada.)

## 9. Pressure–volume relationships in jet-propelled animals

The hydrodynamic work done by locomotor muscles of jet propelled animals (such as squid, and jellyfish) can be calculated by measuring the pressure and volume changes that occur in the central cavity. The area under the loop of the pressure and volume curve produced during the contraction phase is a measure of the energy available to do hydrodynamic work. Pressure can be measured by using a pressure transducer attached to polyethylene tubing, which is inserted into the central cavity. Low cost, low pressure, solid-state silicon pressure sensors can be purchased commercially (Nova Sensor, Fremont, California). These are easily calibrated against known columns of water. Some care must be taken to ensure that air bubbles are not trapped between the transducer and the open end of the tubing. The dynamic response of the system should be measured to ensure that the resonant frequency of the transducer system is about an order of magnitude greater than the frequency of the jet cycle. The dynamic response can be measured from free vibrations generated by a pressure transient applied to the end of the tubing (15). Volume changes can be simultaneously calculated from diameter measurements, which are made with a Video Dimension Analyser (Instruments for Physiology and Medicine, San Diego, California). This instrument produces an electrical signal that is proportional to the distance between two contrast boundaries on the video image. A telemetry system has been recently developed to measure mantle cavity pressure changes in free-swimming squid (16), which allows many aspects of locomotion in these animals to be measured directly.

## 10. Hydromechanical modelling

Data for analysing and modelling animal movement can be derived from high-speed cine-film or video. Body and appendage positions are digitized using frame by frame analysis. For a successful analysis, the framing rate has to be high enough to capture the required details of the movement. Low filming speed tends to underestimate peak accelerations (17). If the film speed is too high, relative errors in the individual measurements of the shorter distances moved by the animal increase. Such measurement errors can be compensated by increasing image magnification. The time-displacement data traced from the film are then smoothed, for which a multitude of methods are available (see also Chapter 3). Gal and Blake (18) chose seven successive points of the time-displacement data as a set, and fitted them with a quadratic polynomial. This generated a single new displacement at the value of the function at the centre (fourth) point. The next set of time-displacement data was chosen by dropping the first data point from the series, and including the eighth point for the next set. It is straightforward to calculate the corresponding velocities and accelerations as first and second derivatives of the smoothed data. Small accelerometers (Kistler Instrument Corp., Amherst, NY) can also be surgically implanted for more reliable estimates of high accelerations (19).

Two methods are widely used to model an animal's movement: *quasi-steady analysis* which gives information about lift and drag, and analysis of the acceleration, which reveals the importance of the acceleration reaction. In quasi-steady analysis, drag and lift are calculated frame by frame, assuming that each represents steady motion. Drag and lift coefficients are calculated from the instantaneous estimates of velocity, assuming that coefficients were previously derived (Section 8). Equations (1) and (3) then give the instantaneous drag and lift. These steady-state forces can be integrated to give the average drag and lift over the cycle of movement.

The quasi-steady approach however ignores inertial forces. The acceleration reaction can be evaluated by measuring the accelerations from the film analysis, or accelerometers. The possible changes in body or appendage geometry due to swimming movements should be accounted for in both the quasi-steady analysis and acceleration reaction analysis. Jointed appendages used in some movements can be modelled as a series of linked geometrical objects, such as cylinders or flat plates. As a first approximation, the inter-action between the segments can often be ignored. A clear and straight-forward treatment of frog swimming, including an analysis of segmented leg movement, is presented by Gal and Blake (18, 20).

## 11. Summary

This chapter introduced the basic theory of fluid dynamics, with some emph-asis on the hydrodynamic forces generated by flow. The importance of the Reynolds number in characterizing several aspects of fluid flow was shown. Dimensional analysis, a useful tool in many biological problems, was intro-duced with an example from fluid dynamics. The last part of the chapter described experimental techniques that can be used to measure many aspects of fluid flow in swimming animals. These techniques have provided informa-tion that has improved our understanding of how animals swim. An example might demonstrate the utility of these concepts. The molluscan squid swims with thrust generated by both jet propulsion and oscillations of fins, the contribution from each system varying with the intensity of the movement. O'Dor (21) used analysis of cine-film, measurement of intramantle pressures, and hydromechanical modelling to partition the contributions of the two mechanisms squid use to generate thrust. The jet cycle itself involves re-peated diameter changes of the muscular mantle that surrounds the viscera. Thrust is generated when the diameter decreases and water contained in a cavity located between the viscera and mantle is ejected through a funnel located at one end of the animal. Refilling is not easily accomplished by contraction of muscles antagonist to the muscles that decrease the diameter, so most of the antagonism is derived from elastic mechanisms. Gosline and Shadwick (22) quantified the importance of the elastic mechanism with measurements of the press–volume relationships of jetting animals. Vogel (7) used models of squid in wind tunnels to show that refilling of the mantle can be assisted by pressure differences between the region of greatest diameter and the refilling apertures located downstream.

## References

1. Denny, M. W. (1988). *Biology and mechanics of the wave-swept environment.* Princeton University Press, Princeton, NJ.
2. Schlichting, H. (1979). *Boundary layer theory* (7th edn). McGraw-Hill, New York.
3. Vogel, S. (1983). *Life in moving fluids—The physical biology of flow.* Princeton University Press, Princeton, NJ.
4. Daniel, T. L. (1985). *J. exp. Biol.,* **119**, 149.
5. Alexander, R. McN. and Goldspink, G. (ed.) (1977). *Mechanics and energetics of animal locomotion.* Chapman & Hall, London.

6. Langhaar, H. L. (1987). *Dimensional analysis and theory of models*. Robert E. Krieger Co., Malabar, Florida.
7. Vogel, S. (1987). *Biol. Bull., 172*, 61.
8. Gal, J. M. and Blake, R. W. (1987). *Can. J. Zool.* **65**(5), 1085.
9. Vogel, S. and LaBarbera, M. (1978). *Bioscience, 28*, 638.
10. Weast, R. C., Astle, M. J., and Beyer, W. H. (ed.) (1988). *CRC handbook of chemistry and physics*. CRC Press, Boca Raton, Florida.
11. Vennard, J. K. and Street, R. L. (1976). *Elementary fluid mechanics*. John Wiley, Toronto, Ontario.
12. LaBarbera, M. and Vogel, S. (1976). *Limnol. Oceanogr., 21*, 750.
13. Beyer, W. H. and Selby, S. M. (ed.) (1976). *CRC standard mathematical tables*. CRC Press, Cleveland, Ohio.
14. Alexander, D. E. (1990). *Biol. Bull., 179*, 186.
15. Glantz, S. A. and Tyberg, J. V. (1979). *Am. J. Physiol., 236*(2), H376.
16. Webber, D. M. and O'Dor, R. K. (1986). *J. exp. Biol., 126*, 205.
17. Harper, D. G. and Blake, R. W. (1989). *Can. J. Zool., 67*, 1929.
18. Gal, J. M. and Blake, R. W. (1988). *J. exp. Biol., 138*, 399.
19. Harper, D. G. and Blake, R. W. (1991). *Can. J. Zool., 155*, 175.
20. Gal, J. M. and Blake, R. W. (1988). *J. exp. Biol., 138*, 413.
21. O'Dor, R. K. (1988). *J. exp. Biol., 137*, 421.
22. Gosline, J. M. and Shadwick, R. E. (1983). *Can. J. Zool., 61*(6), 1421.
23. Denny, M. W. (1983). *Limnol. Oceanogr., 28*(6), 1269.
24. Denny, M. W. (1989). *Can. J. Zool., 67*, 2098.

# A1

# Suppliers of specialist items

For information on suppliers of software, see Chapter 7, Section 6.

**Advanced Mechanical Technology Inc.,** Newton, MA, USA.

**Automatix, Inc.,** 755 Middlesex Turnpike, Billerica, MA, USA. (Fax: 010-1-508-663-5482)

**Baxter Healthcare Corp. (& Baxter Medical Supplies),** Scientific Products Div., 1430 Waukegan Rd., McGraw Park, IL 60085, USA. (Tel: 010-1-708-689-8410)

**Bioengineering Group,** The University of Iowa, 56 M.R.F., Iowa City, IA 52242, USA.

**B. Braun Melsungen AG,** D-3508 Melsungen, Germany.

**BLH Electronics,** 75 Shawmut Rd., Canton, MA 02021, USA. (Fax: 010-1-617-828-1451)

**The Conard Corp.,** PO Box 676, 101 Commerce St., Glastonbury, CT 06033, USA.

**Data Translation, Inc.,** 100 Locke Dr., Marlboro, MA 01752, USA. (Tel: 010-1-508-481-3700)

**Davis and Geck,** Cyanamid of Great Britain Ltd., Gosport, Hampshire, PO13 0AS, UK.

**Eastman Kodak Co.,** 11633 Sorrento Valley Rd., San Diego, CA 92121, USA. (Tel: 010-1-619-481-8182)

**Fischer Scientific, Inc.,** PO Box 121, Waukesha, WI 53187, USA. (Tel: 010-1-414-547-3460)

**Grass Instruments, Inc.,** 101 Old Colony Ave., Quincy, MA 02169, USA. (Tel: 010-1-617-773-0002; Fax: 010-1-617-773-0415)

**GTCO Corp.,** 1055 First St., Rockville, MD 20850, USA. (Tel: 010-1-301-279-9550)

**Hexcel Corp.,** 650 California St., San Francisco, CA 94108, USA.

**Instron Corp.,** 100 Royall St., Canton, MA 02021, USA.

**Instrumentation for Physiology and Medicine, Inc.,** PO Box 19206, San Diego, CA 92119, USA. (Tel: 010-1-619-464-6383)

**Jandel Scientific,** 65 Koch Rd, Corte Madera, CA 94925, USA. (Tel: 010-1-415-924-8640)

**Kistler Instrument Corp.,** 75 John Glenn Dr., Amherst, NY 14120, USA. (Tel: 010-1-716-691-5100)

**Ling Dynamic Systems, Inc.,** 60 Church St., Yalesville, CT 06492, USA. (Tel: 010-1-800-468-6537; 010-1-203-265-7966) or Baldock Rd., Royston, Hertfordshire, UK.

**Matrox Electronics,** 1055 St. Regius Blvd., Dorval, Quebec, Canada H9P 2T4. (Tel: 010-1-514-685-2630)

**MetraByte Corp.,** 440 Myles Standish Blvd., Taunton, MA 02780, USA. (Fax: 010-1-508-880-0179)

**MicroMeasurements,** 5 Forest Park Dr., Farmington, CT 94707, USA. (Tel: 010-1-203-677-2677)

**Millar Instruments, Inc.,** PO Box 230227, Houston, TX 77223, USA. (Tel: 010-1-713-923-9171; Fax: 010-1-713-923-7757)

**Mitutoyo, MTI Corp.,** 18 Essex Rd., Paramus, NJ 07652, USA.

**Monoject Scientific,** 200 Express St., Plainview, NY 11803, USA. (Fax: 010-1-516-349-8719)

**MorphoSys,** C. A. Meacham and T. Duncan, University Herbarium, University of California, Berkeley, CA 94720, USA.

**Motion Analysis Inc.,** 1011 Willagillespie Rd., Eugene, OR 97401, USA. (Tel: 010-1-503-342-3440)

**NAC,** 1011-F West Alameda Ave., Burbank, CA 91506, USA. (Tel: 010-1-800-622-1162)

**Nova Sensor,** 1055 Mission Court, Freemont, CA 94539, USA. (Tel: 010-1-415-490-9100; Fax: 010-1-415-770-0645)

**NTIS,** Research Services Branch of the National Institutes of Health, Rockville Pike, Bethesda, MD, USA.

**Omega Engineering, Inc.,** One Omega Dr., PO Box 4047, Stamford, CT 06907, USA. (Tel: 010-1-203-359-1660)

**Panasonic Corp.,** 425 East Algonquin Rd., Arlington Heights, IL. USA. (Tel: 010-1-708-364-7900)

**Para Tech Coating Company,** Suite #R, 23141 La Cadena Dr., Laguna Hills, CA 92653, USA.

**Peak Performance Technologies Inc.,** 7388 S. Revere Pkwy., #601, Englewood, CO 80112, USA. (Tel: 010-1-303-799-8686)

**Rockware, Inc.,** 4251 Kipling St., Suite 595, Wheat Ridge, CO, USA.

**Reichert-Jung, Cambridge Instruments Inc.,** 111 Dear Lake Rd., Deerfield, IL 60015, USA. (Tel: 010-1-708-405-0123; Fax: 010-1-708-405-0147)

**Summagraphics, Inc.,** 777 State Street Exchange, Fairfield, CT 06430, USA. (Tel: 010-1-203-384-1344)

**Thomas Scientific,** 99 High Hill Rd., Swedesboro, NJ 08085, USA. (Tel: 010-1-800-345-2100; Fax: 010-1-609-467-3087)

**Tokyo Sokki Kenkyujo, Ltd.,** Minami-Ohi 6-Chome, Shinagawa-Ku, Tokyo, 140, Japan. (Fax: Tokyo 03-763-6128)

**Triton Technology, Inc.,** 4616 Santa Fe St., San Diego, CA 92109, USA. (010-1-619-272-1251)

**TSI, Inc.,** 500 Cardigan Rd., PO Box 64394, St. Paul, MN 55164, USA. (Tel: 010-1-612-483-0900; Fax: 010-1-612-481-1220)
**Viking Technology, Inc.,** 1620 Berryessa Rd., San Jose, CA 95133, USA.

# Index